9000 Mathematics Games to Sharpen your Intellect and give your the Power of Arithmetic Thinking

By: Davin Rhoades

Mathematics can be Tricky

Our biggest strength, our greatest weapon, our most valuable asset: is our mind. And many of us never realize this, and spend our lives without harnessing even a tiny fraction of the immense power that our brains possess.

However, it is always possible, irrespective of your age, to make a fresh start, and to get your mind in the right shape to conquer every problem that is posed to it.

This book has a series of carefully chosen exercises that will improve the mathematical faculties of your brain.

I have deliberately chosen the material in this book such that you will not need many prerequisites to understand the questions and to solve them correctly: for me, it's all about the basic digits, since they can produce great complexity. Therefore, you will only need to know basic arithmetic, and possibly the PEMDAS order of operations, to solve everything in this book.

However, it's not always easy: if you want to get a perfect score, you will have to spend hundreds, if not thousands, of hours trying to solve some of the harder problems in this book.

Fortunately, I have spent the time to give you all the solutions, so that you will not be left alone for the exercises that are too hard to complete unaided.

Go on, take your first few steps to becoming a Mind Warrior right now!

Table of Contents

The 24 Challenge

This is deceptively simple!

You are given four numbers. Using those, and the four arithmetic operations, you have to reach the number 24.

For example, if you were given the digits:

1, 4, 4, and 5:

We can see that

1 x 4 x 5 + 4 = 20 + 4 = 24.

Can you solve these two:

A. 7, 3, 1, 1

B. 8, 1, 1, 1

1 1 3 7

$(7+1)×3×1$

$(7-1)×(3+1)$

1, 1, 1, 8

$(1+1+1)×8 = 24$

What Comes Next?

Simple: Each of these features a different type of arithmetic sequence / series, and you have to figure out the next member of the sequence.

Good luck!

{ 1 }

Solve this: what comes next in this sequence?:

148, 159, 170, 181, 192, ...?

||||

{ 2 }

Which number fits in the next spot in this sequence?:

4.1, 12.3, 36.9, 110.7, 332.1, ...?

||||

{ 3 }

Which number fits in the next spot in this sequence?:

160, 481, 1444, 4333, 13000, ...?

||||

Check Your Answers: What Comes Next?

{ 1 }

Solve this: what comes next in this sequence?:

148, 159, 170, 181, 192, ...?

||||

By taking the differences between adjacent terms, we can see that this is an Arithmetic Progression, with common difference = 11. Therefore, the sixth term can be calculated by adding the common difference to the fifth term. It is equal to 192 + 11 = 203.

|||||||||

{ 2 }

Which number fits in the next spot in this sequence?:

4.1, 12.3, 36.9, 110.7, 332.1, ...?

||||

By taking the ratios of adjacent terms, we can see that this is a Geometric Progression, with common ratio = 3. Therefore, the sixth term can be calculated by multiplying the fifth term with the common ratio. It is equal to 332.1 x 3 = 996.3.

|||||||||

{ 3 }

Which number fits in the next spot in this sequence?:

160, 481, 1444, 4333, 13000, ...?

||||

By observation and trial and error, we find that each term in this

sequence is obtained by multiplying the previous term by 3 and adding 1.

Therefore, the sixth term = (3 x 13000) + 1 = 39001.

||||||||

Five 5s

Easy to explain; hard to achieve!

The goal: to use exactly five 5s (and no other digits) along with the four arithmetic operations, brackets, decimal points, and factorial signs, to end up with each positive integer, one by one.

For example, we can see that:

$$14 = 5 + 5 + 5 - (5/5)$$

How many of the numbers from 1 to 10 can you make using five 5s and the operations listed above? Take half an hour, and then head over to the answers to see how you did!

$$1 = \frac{5 \times 5 \times 5 - 5}{5!}$$

$$3 = \frac{((55 / 5) - 5)}{.5}$$

$$4 = (5 / (5 / 5)) - (5 / 5)$$

$$5 = 5 / (55 / 55)$$

$$6 = (5 - (5 / 5) - (5 / 5))\,!$$

$$7 = 5 + 5/5 + 5/5$$

$$8 = (5 + (55 / 5)) * .5$$

$$9 = \sqrt{5} \times \sqrt{5} + 5 - \dfrac{5}{5}$$

$$10 = \frac{5 \times 5 + 5 \times 5}{5}$$

Series-ly: Part 1

{ 1 }

Which number fits in the next spot in this sequence?:

180, 361, 723, 1447, 2895, ...?

IIII

{ 2 }

Which number fits in the next spot in this sequence?:

136, 409, 1228, 3685, 11056, ...?

IIII

{ 3 }

Solve this: what comes next in this sequence?:

105, 147, 189, 231, 273, ...?

IIII

{ 4 }

Which number fits in the next spot in this sequence?:

211, 845, 3381, 13525, 54101, ...?

IIII

{ 5 }

Solve this: what comes next in this sequence?:

196, 222, 248, 274, 300, ...?

||||

{ 6 }

Which number fits in the next spot in this sequence?:

9, 27, 81, 243, 729, ...?

||||

{ 7 }

Solve this: what comes next in this sequence?:

275, 288, 301, 314, 327, ...?

||||

{ 8 }

Solve this: what comes next in this sequence?:

206, 243, 280, 317, 354, ...?

||||

{ 9 }

Solve this: what comes next in this sequence?:

242, 265, 288, 311, 334, ...?

||||

{ 10 }

Which number fits in the next spot in this sequence?:

188, 565, 1696, 5089, 15268, ...?

||||

Check Your Answers: Series-ly: Part 1

{ 1 }

Which number fits in the next spot in this sequence?:

180, 361, 723, 1447, 2895, ...?

||||

By observation and trial and error, we find that each term in this sequence is obtained by multiplying the previous term by 2 and adding 1. Therefore, the sixth term = (2 x 2895) + 1 = 5791.

||||||||

{ 2 }

Which number fits in the next spot in this sequence?:

136, 409, 1228, 3685, 11056, ...?

||||

By observation and trial and error, we find that each term in this sequence is obtained by multiplying the previous term by 3 and adding 1. Therefore, the sixth term = (3 x 11056) + 1 = 33169.

||||||||

{ 3 }

Solve this: what comes next in this sequence?:

105, 147, 189, 231, 273, ...?

||||

By taking the differences between adjacent terms, we can see that this is an Arithmetic Progression, with common difference = 42. Therefore, the sixth term can be calculated by adding the common difference to the fifth term. It is equal to 273 + 42 = 315.

|||||||||

{ 4 }

Which number fits in the next spot in this sequence?:

211, 845, 3381, 13525, 54101, ...?

||||

By observation and trial and error, we find that each term in this sequence is obtained by multiplying the previous term by 4 and adding 1. Therefore, the sixth term = (4 x 54101) + 1 = 216405.

||||||||

{ 5 }

Solve this: what comes next in this sequence?:

196, 222, 248, 274, 300, ...?

||||

By taking the differences between adjacent terms, we can see that this is an Arithmetic Progression, with common difference = 26. Therefore, the

sixth term can be calculated by adding the common difference to the fifth

term. It is equal to 300 + 26 = 326.

|||||||||

{ 6 }

Which number fits in the next spot in this sequence?:

9, 27, 81, 243, 729, ...?

||||

By taking the ratios of adjacent terms, we can see that this is a

Geometric Progression, with common ratio = 3. Therefore, the sixth term

can be calculated by multiplying the fifth term with the common ratio. It is

equal to 729 x 3 = 2187.

|||||||||

{ 7 }

Solve this: what comes next in this sequence?:

275, 288, 301, 314, 327, ...?

||||

By taking the differences between adjacent terms, we can see that this is

an Arithmetic Progression, with common difference = 13. Therefore, the

sixth term can be calculated by adding the common difference to the fifth

term. It is equal to 327 + 13 = 340.

|||||||||

{ 8 }

Solve this: what comes next in this sequence?:

206, 243, 280, 317, 354, ...?

||||

By taking the differences between adjacent terms, we can see that this is an Arithmetic Progression, with common difference = 37. Therefore, the sixth term can be calculated by adding the common difference to the fifth term. It is equal to 354 + 37 = 391.

|||||||||

{ 9 }

Solve this: what comes next in this sequence?:

242, 265, 288, 311, 334, ...?

||||

By taking the differences between adjacent terms, we can see that this is an Arithmetic Progression, with common difference = 23. Therefore, the sixth term can be calculated by adding the common difference to the fifth term. It is equal to 334 + 23 = 357.

|||||||||

{ 10 }

Which number fits in the next spot in this sequence?:

188, 565, 1696, 5089, 15268, ...?

||||

By observation and trial and error, we find that each term in this
sequence is obtained by multiplying the previous term by 3 and adding 1.
Therefore, the sixth term = (3 x 15268) + 1 = 45805.

|||||||||

Series-ly: Part 2

{ 11 }

Solve this: what comes next in this sequence?:

145, 167, 189, 211, 233, ...?

||||

{ 12 }

Solve this: what comes next in this sequence?:

219, 268, 317, 366, 415, ...?

||||

{ 13 }

Which number fits in the next spot in this sequence?:

8.6, 34.4, 137.6, 550.4, 2201.6, ...?

||||

{ 14 }

Which number fits in the next spot in this sequence?:

6.6, 26.4, 105.6, 422.4, 1689.6, ...?

||||

{ 15 }

Which number fits in the next spot in this sequence?:

6.1, 24.4, 97.6, 390.4, 1561.6, ...?

||||

{ 16 }

Which number fits in the next spot in this sequence?:

2.7, 8.1, 24.3, 72.9, 218.7, ...?

||||

{ 17 }

Which number fits in the next spot in this sequence?:

107, 322, 967, 2902, 8707, ...?

||||

{ 18 }

Solve this: what comes next in this sequence?:

284, 313, 342, 371, 400, ...?

||||

{ 19 }

Solve this: what comes next in this sequence?:

171, 199, 227, 255, 283, ...?

||||

{ 20 }

Which number fits in the next spot in this sequence?:

211, 423, 847, 1695, 3391, ...?

||||

{ 11 }

Solve this: what comes next in this sequence?:

145, 167, 189, 211, 233, ...?

||||

By taking the differences between adjacent terms, we can see that this is an Arithmetic Progression, with common difference = 22. Therefore, the sixth term can be calculated by adding the common difference to the fifth term. It is equal to 233 + 22 = 255.

||||||||

{ 12 }

Solve this: what comes next in this sequence?:

219, 268, 317, 366, 415, ...?

||||

By taking the differences between adjacent terms, we can see that this is an Arithmetic Progression, with common difference = 49. Therefore, the sixth term can be calculated by adding the common difference to the fifth term. It is equal to 415 + 49 = 464.

||||||||

{ 13 }

Which number fits in the next spot in this sequence?:

8.6, 34.4, 137.6, 550.4, 2201.6, ...?

||||

By taking the ratios of adjacent terms, we can see that this is a

Geometric Progression, with common ratio = 4. Therefore, the sixth term

can be calculated by multiplying the fifth term with the common ratio. It is

equal to 2201.6 x 4 = 8806.4.

|||||||||

{ 14 }

Which number fits in the next spot in this sequence?:

6.6, 26.4, 105.6, 422.4, 1689.6, ...?

||||

By taking the ratios of adjacent terms, we can see that this is a

Geometric Progression, with common ratio = 4. Therefore, the sixth term

can be calculated by multiplying the fifth term with the common ratio. It is

equal to 1689.6 x 4 = 6758.4.

|||||||||

{ 15 }

Which number fits in the next spot in this sequence?:

6.1, 24.4, 97.6, 390.4, 1561.6, ...?

||||

By taking the ratios of adjacent terms, we can see that this is a

Geometric Progression, with common ratio = 4. Therefore, the sixth term

can be calculated by multiplying the fifth term with the common ratio. It is

equal to 1561.6 x 4 = 6246.4.

|||||||||

{ 16 }

Which number fits in the next spot in this sequence?:

2.7, 8.1, 24.3, 72.9, 218.7, ...?

||||

By taking the ratios of adjacent terms, we can see that this is a

Geometric Progression, with common ratio = 3. Therefore, the sixth term

can be calculated by multiplying the fifth term with the common ratio. It is

equal to 218.7 x 3 = 656.100000000001.

|||||||||

{ 17 }

Which number fits in the next spot in this sequence?:

107, 322, 967, 2902, 8707, ...?

||||

By observation and trial and error, we find that each term in this

sequence is obtained by multiplying the previous term by 3 and adding 1.

Therefore, the sixth term = (3 x 8707) + 1 = 26122.

|||||||||

{ 18 }

Solve this: what comes next in this sequence?:

284, 313, 342, 371, 400, ...?

||||

By taking the differences between adjacent terms, we can see that this is an Arithmetic Progression, with common difference = 29. Therefore, the sixth term can be calculated by adding the common difference to the fifth term. It is equal to 400 + 29 = 429.

|||||||||

{ 19 }

Solve this: what comes next in this sequence?:

171, 199, 227, 255, 283, ...?

||||

By taking the differences between adjacent terms, we can see that this is an Arithmetic Progression, with common difference = 28. Therefore, the sixth term can be calculated by adding the common difference to the fifth term. It is equal to 283 + 28 = 311.

|||||||||

{ 20 }

Which number fits in the next spot in this sequence?:

211, 423, 847, 1695, 3391, ...?

||||

By observation and trial and error, we find that each term in this

sequence is obtained by multiplying the previous term by 2 and adding 1.

Therefore, the sixth term = (2 x 3391) + 1 = 6783.

|||||||||

Series-ly: Part 3

{ 21 }

Which number fits in the next spot in this sequence?:

160, 641, 2565, 10261, 41045, ...?

||||

{ 22 }

Which number fits in the next spot in this sequence?:

93, 280, 841, 2524, 7573, ...?

||||

{ 23 }

Which number fits in the next spot in this sequence?:

145, 291, 583, 1167, 2335, ...?

||||

{ 24 }

Which number fits in the next spot in this sequence?:

189, 568, 1705, 5116, 15349, ...?

||||

{ 25 }

Solve this: what comes next in this sequence?:

144, 180, 216, 252, 288, ...?

||||

{ 26 }

Which number fits in the next spot in this sequence?:

9.6, 28.8, 86.4, 259.2, 777.6, ...?

||||

{ 27 }

Solve this: what comes next in this sequence?:

150, 176, 202, 228, 254, ...?

||||

{ 28 }

Which number fits in the next spot in this sequence?:

3.2, 9.6, 28.8, 86.4, 259.2, ...?

||||

{ 29 }

Which number fits in the next spot in this sequence?:

1.1, 4.4, 17.6, 70.4, 281.6, ...?

||||

{ 30 }

Which number fits in the next spot in this sequence?:

7.4, 29.6, 118.4, 473.6, 1894.4, ...?

||||

{ 31 }

Which number fits in the next spot in this sequence?:

238, 715, 2146, 6439, 19318, ...?

||||

{ 32 }

Which number fits in the next spot in this sequence?:

2, 8, 32, 128, 512, ...?

||||

{ 33 }

Which number fits in the next spot in this sequence?:

6.9, 20.7, 62.1, 186.3, 558.9, ...?

||||

{ 34 }

Solve this: what comes next in this sequence?:

107, 133, 159, 185, 211, ...?

||||

{ 35 }

Which number fits in the next spot in this sequence?:

7.7, 30.8, 123.2, 492.8, 1971.2, ...?

||||

{ 36 }

Solve this: what comes next in this sequence?:

298, 320, 342, 364, 386, ...?

||||

{ 37 }

Which number fits in the next spot in this sequence?:

8.8, 17.6, 35.2, 70.4, 140.8, ...?

||||

{ 38 }

Which number fits in the next spot in this sequence?:

7.8, 23.4, 70.2, 210.6, 631.8, ...?

||||

{ 39 }

Which number fits in the next spot in this sequence?:

220, 661, 1984, 5953, 17860, ...?

||||

{ 40 }

Which number fits in the next spot in this sequence?:

164, 329, 659, 1319, 2639, ...?

||||

{ 41 }

Solve this: what comes next in this sequence?:

268, 315, 362, 409, 456, ...?

||||

{ 42 }

Which number fits in the next spot in this sequence?:

124, 373, 1120, 3361, 10084, ...?

||||

{ 43 }

Which number fits in the next spot in this sequence?:

4.1, 12.3, 36.9, 110.7, 332.1, ...?

||||

{ 44 }

Solve this: what comes next in this sequence?:

258, 291, 324, 357, 390, ...?

||||

{ 45 }

Which number fits in the next spot in this sequence?:

9.1, 18.2, 36.4, 72.8, 145.6, ...?

||||

{ 46 }

Which number fits in the next spot in this sequence?:

99, 199, 399, 799, 1599, ...?

||||

{ 47 }

Solve this: what comes next in this sequence?:

153, 203, 253, 303, 353, ...?

||||

{ 48 }

Which number fits in the next spot in this sequence?:

5, 15, 45, 135, 405, ...?

||||

{ 49 }

Which number fits in the next spot in this sequence?:

6.7, 13.4, 26.8, 53.6, 107.2, ...?

||||

{ 50 }

Which number fits in the next spot in this sequence?:

8.2, 24.6, 73.8, 221.4, 664.2, ...?

||||

{ 51 }

Which number fits in the next spot in this sequence?:

5.6, 22.4, 89.6, 358.4, 1433.6, ...?

||||

{ 52 }

Solve this: what comes next in this sequence?:

238, 277, 316, 355, 394, ...?

||||

{ 53 }

Solve this: what comes next in this sequence?:

153, 191, 229, 267, 305, ...?

||||

{ 54 }

Solve this: what comes next in this sequence?:

214, 224, 234, 244, 254, ...?

||||

{ 55 }

Which number fits in the next spot in this sequence?:

1.2, 4.8, 19.2, 76.8, 307.2, ...?

||||

{ 56 }

Which number fits in the next spot in this sequence?:

4.6, 9.2, 18.4, 36.8, 73.6, ...?

||||

{ 57 }

Solve this: what comes next in this sequence?:

171, 194, 217, 240, 263, ...?

||||

{ 58 }

Which number fits in the next spot in this sequence?:

155, 466, 1399, 4198, 12595, ...?

||||

{ 59 }

Which number fits in the next spot in this sequence?:

6.1, 18.3, 54.9, 164.7, 494.1, ...?

||||

{ 60 }

Which number fits in the next spot in this sequence?:

257, 1029, 4117, 16469, 65877, ...?

||||

{ 61 }

Which number fits in the next spot in this sequence?:

9.7, 19.4, 38.8, 77.6, 155.2, ...?

||||

{ 62 }

Which number fits in the next spot in this sequence?:

124, 497, 1989, 7957, 31829, ...?

||||

{ 63 }

Solve this: what comes next in this sequence?:

191, 210, 229, 248, 267, ...?

||||

{ 64 }

Solve this: what comes next in this sequence?:

192, 226, 260, 294, 328, ...?

||||

{ 65 }

Which number fits in the next spot in this sequence?:

3.7, 7.4, 14.8, 29.6, 59.2, ...?

||||

{ 66 }

Which number fits in the next spot in this sequence?:

110, 221, 443, 887, 1775, ...?

||||

{ 67 }

Which number fits in the next spot in this sequence?:

5.2, 15.6, 46.8, 140.4, 421.2, ...?

||||

{ 68 }

Solve this: what comes next in this sequence?:

294, 339, 384, 429, 474, ...?

||||

{ 69 }

Which number fits in the next spot in this sequence?:

6.6, 19.8, 59.4, 178.2, 534.6, ...?

||||

{ 70 }

Which number fits in the next spot in this sequence?:

131, 394, 1183, 3550, 10651, ...?

||||

{ 71 }

Which number fits in the next spot in this sequence?:

8.3, 33.2, 132.8, 531.2, 2124.8, ...?

||||

{ 72 }

Solve this: what comes next in this sequence?:

230, 262, 294, 326, 358, ...?

||||

{ 73 }

Which number fits in the next spot in this sequence?:

223, 670, 2011, 6034, 18103, ...?

||||

{ 74 }

Which number fits in the next spot in this sequence?:

1.6, 6.4, 25.6, 102.4, 409.6, ...?

||||

{ 75 }

Solve this: what comes next in this sequence?:

173, 220, 267, 314, 361, ...?

||||

{ 76 }

Which number fits in the next spot in this sequence?:

278, 835, 2506, 7519, 22558, ...?

||||

{ 77 }

Which number fits in the next spot in this sequence?:

157, 472, 1417, 4252, 12757, ...?

||||

{ 78 }

Solve this: what comes next in this sequence?:

280, 297, 314, 331, 348, ...?

||||

{ 79 }

Which number fits in the next spot in this sequence?:

223, 447, 895, 1791, 3583, ...?

||||

{ 80 }

Which number fits in the next spot in this sequence?:

5, 20, 80, 320, 1280, ...?

||||

{ 81 }

Solve this: what comes next in this sequence?:

100, 141, 182, 223, 264, ...?

||||

{ 82 }

Solve this: what comes next in this sequence?:

185, 233, 281, 329, 377, ...?

||||

{ 83 }

Which number fits in the next spot in this sequence?:

4.9, 14.7, 44.1, 132.3, 396.9, ...?

||||

{ 84 }

Which number fits in the next spot in this sequence?:

266, 1065, 4261, 17045, 68181, ...?

||||

{ 85 }

Solve this: what comes next in this sequence?:

100, 137, 174, 211, 248, ...?

||||

{ 86 }

Solve this: what comes next in this sequence?:

287, 317, 347, 377, 407, ...?

||||

{ 87 }

Which number fits in the next spot in this sequence?:

4.6, 13.8, 41.4, 124.2, 372.6, ...?

||||

{ 88 }

Solve this: what comes next in this sequence?:

132, 162, 192, 222, 252, ...?

||||

{ 89 }

Which number fits in the next spot in this sequence?:

276, 553, 1107, 2215, 4431, ...?

||||

{ 90 }

Which number fits in the next spot in this sequence?:

5.5, 16.5, 49.5, 148.5, 445.5, ...?

||||

{ 91 }

Solve this: what comes next in this sequence?:

296, 318, 340, 362, 384, ...?

||||

{ 92 }

Solve this: what comes next in this sequence?:

188, 219, 250, 281, 312, ...?

||||

{ 93 }

Which number fits in the next spot in this sequence?:

7.3, 14.6, 29.2, 58.4, 116.8, ...?

||||

{ 94 }

Which number fits in the next spot in this sequence?:

1.6, 3.2, 6.4, 12.8, 25.6, ...?

||||

{ 95 }

Solve this: what comes next in this sequence?:

280, 329, 378, 427, 476, ...?

||||

{ 96 }

Which number fits in the next spot in this sequence?:

1.5, 6, 24, 96, 384, ...?

||||

{ 97 }

Solve this: what comes next in this sequence?:

211, 259, 307, 355, 403, ...?

||||

{ 98 }

Which number fits in the next spot in this sequence?:

162, 487, 1462, 4387, 13162, ...?

||||

{ 99 }

Which number fits in the next spot in this sequence?:

205, 616, 1849, 5548, 16645, ...?

||||

{ 100 }

Solve this: what comes next in this sequence?:

250, 293, 336, 379, 422, ...?

||||

{ 101 }

Which number fits in the next spot in this sequence?:

6.7, 26.8, 107.2, 428.8, 1715.2, ...?

||||

{ 102 }

Which number fits in the next spot in this sequence?:

249, 748, 2245, 6736, 20209, ...?

||||

{ 103 }

Which number fits in the next spot in this sequence?:

7, 21, 63, 189, 567, ...?

||||

{ 104 }

Solve this: what comes next in this sequence?:

165, 211, 257, 303, 349, ...?

||||

{ 105 }

Solve this: what comes next in this sequence?:

279, 292, 305, 318, 331, ...?

||||

{ 106 }

Which number fits in the next spot in this sequence?:

193, 773, 3093, 12373, 49493, ...?

||||

{ 107 }

Which number fits in the next spot in this sequence?:

159, 319, 639, 1279, 2559, ...?

||||

{ 108 }

Which number fits in the next spot in this sequence?:

5.3, 15.9, 47.7, 143.1, 429.3, ...?

||||

{ 109 }

Solve this: what comes next in this sequence?:

239, 265, 291, 317, 343, ...?

||||

{ 110 }

Which number fits in the next spot in this sequence?:

6.2, 12.4, 24.8, 49.6, 99.2, ...?

||||

{ 111 }

Which number fits in the next spot in this sequence?:

7, 21, 63, 189, 567, ...?

||||

{ 112 }

Which number fits in the next spot in this sequence?:

8.8, 17.6, 35.2, 70.4, 140.8, ...?

||||

{ 113 }

Which number fits in the next spot in this sequence?:

216, 649, 1948, 5845, 17536, ...?

||||

{ 114 }

Which number fits in the next spot in this sequence?:

5.8, 23.2, 92.8, 371.2, 1484.8, ...?

||||

{ 115 }

Which number fits in the next spot in this sequence?:

180, 361, 723, 1447, 2895, ...?

||||

{ 116 }

Which number fits in the next spot in this sequence?:

9.6, 19.2, 38.4, 76.8, 153.6, ...?

||||

{ 117 }

Which number fits in the next spot in this sequence?:

4.3, 8.6, 17.2, 34.4, 68.8, ...?

||||

{ 118 }

Solve this: what comes next in this sequence?:

254, 266, 278, 290, 302, ...?

||||

{ 119 }

Which number fits in the next spot in this sequence?:

5.8, 23.2, 92.8, 371.2, 1484.8, ...?

||||

{ 120 }

Which number fits in the next spot in this sequence?:

298, 597, 1195, 2391, 4783, ...?

||||

{ 121 }

Which number fits in the next spot in this sequence?:

204, 817, 3269, 13077, 52309, ...?

||||

{ 122 }

Which number fits in the next spot in this sequence?:

9.6, 28.8, 86.4, 259.2, 777.6, ...?

||||

{ 123 }

Which number fits in the next spot in this sequence?:

252, 505, 1011, 2023, 4047, ...?

||||

{ 124 }

Solve this: what comes next in this sequence?:

148, 188, 228, 268, 308, ...?

||||

{ 125 }

Which number fits in the next spot in this sequence?:

1.6, 6.4, 25.6, 102.4, 409.6, ...?

||||

{ 126 }

Solve this: what comes next in this sequence?:

221, 251, 281, 311, 341, ...?

||||

{ 127 }

Which number fits in the next spot in this sequence?:

3.9, 15.6, 62.4, 249.6, 998.4, ...?

||||

{ 128 }

Solve this: what comes next in this sequence?:

197, 208, 219, 230, 241, ...?

||||

{ 129 }

Which number fits in the next spot in this sequence?:

210, 841, 3365, 13461, 53845, ...?

||||

{ 130 }

Which number fits in the next spot in this sequence?:

171, 343, 687, 1375, 2751, ...?

||||

{ 131 }

Solve this: what comes next in this sequence?:

95, 130, 165, 200, 235, ...?

||||

{ 132 }

Solve this: what comes next in this sequence?:

270, 306, 342, 378, 414, ...?

||||

{ 133 }

Which number fits in the next spot in this sequence?:

191, 383, 767, 1535, 3071, ...?

||||

{ 134 }

Solve this: what comes next in this sequence?:

116, 154, 192, 230, 268, ...?

||||

{ 135 }

Solve this: what comes next in this sequence?:

139, 173, 207, 241, 275, ...?

||||

{ 136 }

Which number fits in the next spot in this sequence?:

299, 898, 2695, 8086, 24259, ...?

||||

{ 137 }

Which number fits in the next spot in this sequence?:

222, 667, 2002, 6007, 18022, ...?

||||

{ 138 }

Solve this: what comes next in this sequence?:

177, 222, 267, 312, 357, ...?

||||

{ 139 }

Which number fits in the next spot in this sequence?:

2.5, 5, 10, 20, 40, ...?

||||

{ 140 }

Which number fits in the next spot in this sequence?:

1.3, 5.2, 20.8, 83.2, 332.8, ...?

||||

{ 141 }

Which number fits in the next spot in this sequence?:

296, 593, 1187, 2375, 4751, ...?

||||

{ 142 }

Which number fits in the next spot in this sequence?:

166, 499, 1498, 4495, 13486, ...?

||||

{ 143 }

Solve this: what comes next in this sequence?:

119, 156, 193, 230, 267, ...?

||||

{ 144 }

Solve this: what comes next in this sequence?:

202, 249, 296, 343, 390, ...?

||||

{ 145 }

Which number fits in the next spot in this sequence?:

8.4, 25.2, 75.6, 226.8, 680.4, ...?

||||

{ 146 }

Which number fits in the next spot in this sequence?:

9.3, 27.9, 83.7, 251.1, 753.3, ...?

||||

{ 147 }

Solve this: what comes next in this sequence?:

170, 191, 212, 233, 254, ...?

||||

{ 148 }

Which number fits in the next spot in this sequence?:

4.5, 18, 72, 288, 1152, ...?

||||

{ 149 }

Which number fits in the next spot in this sequence?:

3.7, 7.4, 14.8, 29.6, 59.2, ...?

||||

{ 150 }

Solve this: what comes next in this sequence?:

208, 222, 236, 250, 264, ...?

||||

{ 151 }

Which number fits in the next spot in this sequence?:

7.3, 29.2, 116.8, 467.2, 1868.8, ...?

||||

{ 152 }

Which number fits in the next spot in this sequence?:

6.6, 19.8, 59.4, 178.2, 534.6, ...?

||||

{ 153 }

Which number fits in the next spot in this sequence?:

5.6, 11.2, 22.4, 44.8, 89.6, ...?

||||

{ 154 }

Solve this: what comes next in this sequence?:

207, 243, 279, 315, 351, ...?

||||

{ 155 }

Which number fits in the next spot in this sequence?:

5.2, 15.6, 46.8, 140.4, 421.2, ...?

||||

{ 156 }

Which number fits in the next spot in this sequence?:

256, 513, 1027, 2055, 4111, ...?

||||

{ 157 }

Which number fits in the next spot in this sequence?:

6.3, 12.6, 25.2, 50.4, 100.8, ...?

||||

{ 158 }

Solve this: what comes next in this sequence?:

205, 237, 269, 301, 333, ...?

||||

{ 159 }

Solve this: what comes next in this sequence?:

161, 191, 221, 251, 281, ...?

||||

{ 160 }

Which number fits in the next spot in this sequence?:

162, 487, 1462, 4387, 13162, ...?

||||

{ 161 }

Which number fits in the next spot in this sequence?:

229, 459, 919, 1839, 3679, ...?

||||

{ 162 }

Which number fits in the next spot in this sequence?:

274, 823, 2470, 7411, 22234, ...?

||||

{ 163 }

Which number fits in the next spot in this sequence?:

1.2, 2.4, 4.8, 9.6, 19.2, ...?

||||

{ 164 }

Which number fits in the next spot in this sequence?:

148, 593, 2373, 9493, 37973, ...?

||||

{ 165 }

Solve this: what comes next in this sequence?:

121, 131, 141, 151, 161, ...?

||||

{ 166 }

Which number fits in the next spot in this sequence?:

155, 466, 1399, 4198, 12595, ...?

||||

{ 167 }

Which number fits in the next spot in this sequence?:

6.3, 18.9, 56.7, 170.1, 510.3, ...?

||||

{ 168 }

Which number fits in the next spot in this sequence?:

2.1, 6.3, 18.9, 56.7, 170.1, ...?

||||

{ 169 }

Which number fits in the next spot in this sequence?:

277, 832, 2497, 7492, 22477, ...?

||||

{ 170 }

Which number fits in the next spot in this sequence?:

104, 417, 1669, 6677, 26709, ...?

||||

{ 171 }

Which number fits in the next spot in this sequence?:

9.9, 29.7, 89.1, 267.3, 801.9, ...?

||||

{ 172 }

Which number fits in the next spot in this sequence?:

3.9, 7.8, 15.6, 31.2, 62.4, ...?

||||

{ 173 }

Which number fits in the next spot in this sequence?:

209, 837, 3349, 13397, 53589, ...?

||||

{ 174 }

Solve this: what comes next in this sequence?:

241, 260, 279, 298, 317, ...?

||||

{ 175 }

Which number fits in the next spot in this sequence?:

3.2, 12.8, 51.2, 204.8, 819.2, ...?

||||

{ 176 }

Solve this: what comes next in this sequence?:

237, 274, 311, 348, 385, ...?

||||

{ 177 }

Which number fits in the next spot in this sequence?:

1.5, 6, 24, 96, 384, ...?

||||

{ 178 }

Which number fits in the next spot in this sequence?:

194, 583, 1750, 5251, 15754, ...?

||||

{ 179 }

Solve this: what comes next in this sequence?:

158, 185, 212, 239, 266, ...?

||||

{ 180 }

Which number fits in the next spot in this sequence?:

3, 12, 48, 192, 768, ...?

||||

{ 181 }

Which number fits in the next spot in this sequence?:

295, 886, 2659, 7978, 23935, ...?

||||

{ 182 }

Which number fits in the next spot in this sequence?:

237, 712, 2137, 6412, 19237, ...?

||||

{ 183 }

Which number fits in the next spot in this sequence?:

6.3, 12.6, 25.2, 50.4, 100.8, ...?

||||

{ 184 }

Solve this: what comes next in this sequence?:

130, 180, 230, 280, 330, ...?

||||

{ 185 }

Solve this: what comes next in this sequence?:

163, 205, 247, 289, 331, ...?

||||

{ 186 }

Solve this: what comes next in this sequence?:

173, 215, 257, 299, 341, ...?

||||

{ 187 }

Which number fits in the next spot in this sequence?:

90, 181, 363, 727, 1455, ...?

||||

{ 188 }

Which number fits in the next spot in this sequence?:

6.3, 12.6, 25.2, 50.4, 100.8, ...?

||||

{ 189 }

Solve this: what comes next in this sequence?:

248, 295, 342, 389, 436, ...?

||||

{ 190 }

Which number fits in the next spot in this sequence?:

171, 343, 687, 1375, 2751, ...?

||||

{ 191 }

Which number fits in the next spot in this sequence?:

235, 706, 2119, 6358, 19075, ...?

||||

{ 192 }

Which number fits in the next spot in this sequence?:

3.4, 6.8, 13.6, 27.2, 54.4, ...?

||||

{ 193 }

Which number fits in the next spot in this sequence?:

118, 473, 1893, 7573, 30293, ...?

||||

{ 194 }

Which number fits in the next spot in this sequence?:

4.8, 14.4, 43.2, 129.6, 388.8, ...?

||||

{ 195 }

Which number fits in the next spot in this sequence?:

5, 10, 20, 40, 80, ...?

||||

{ 196 }

Which number fits in the next spot in this sequence?:

105, 316, 949, 2848, 8545, ...?

||||

{ 197 }

Solve this: what comes next in this sequence?:

286, 332, 378, 424, 470, ...?

||||

{ 198 }

Which number fits in the next spot in this sequence?:

93, 280, 841, 2524, 7573, ...?

||||

{ 199 }

Solve this: what comes next in this sequence?:

135, 145, 155, 165, 175, ...?

||||

{ 200 }

Which number fits in the next spot in this sequence?:

8.1, 32.4, 129.6, 518.4, 2073.6, ...?

||||

{ 201 }

Solve this: what comes next in this sequence?:

106, 134, 162, 190, 218, ...?

||||

{ 202 }

Which number fits in the next spot in this sequence?:

9.1, 27.3, 81.9, 245.7, 737.1, ...?

||||

{ 203 }

Solve this: what comes next in this sequence?:

120, 135, 150, 165, 180, ...?

||||

{ 204 }

Which number fits in the next spot in this sequence?:

282, 565, 1131, 2263, 4527, ...?

||||

{ 205 }

Which number fits in the next spot in this sequence?:

9.6, 19.2, 38.4, 76.8, 153.6, ...?

||||

{ 206 }

Solve this: what comes next in this sequence?:

235, 278, 321, 364, 407, ...?

||||

{ 207 }

Which number fits in the next spot in this sequence?:

5.8, 11.6, 23.2, 46.4, 92.8000000000001, ...?

||||

{ 208 }

Which number fits in the next spot in this sequence?:

237, 949, 3797, 15189, 60757, ...?

||||

{ 209 }

Which number fits in the next spot in this sequence?:

3.5, 7, 14, 28, 56, ...?

||||

{ 210 }

Which number fits in the next spot in this sequence?:

117, 469, 1877, 7509, 30037, ...?

||||

{ 211 }

Solve this: what comes next in this sequence?:

142, 180, 218, 256, 294, ...?

||||

{ 212 }

Solve this: what comes next in this sequence?:

281, 322, 363, 404, 445, ...?

||||

{ 213 }

Solve this: what comes next in this sequence?:

228, 262, 296, 330, 364, ...?

||||

{ 214 }

Which number fits in the next spot in this sequence?:

2.2, 6.6, 19.8, 59.4, 178.2, ...?

||||

{ 215 }

Which number fits in the next spot in this sequence?:

286, 1145, 4581, 18325, 73301, ...?

||||

{ 216 }

Solve this: what comes next in this sequence?:

232, 271, 310, 349, 388, ...?

||||

{ 217 }

Which number fits in the next spot in this sequence?:

4.5, 9, 18, 36, 72, ...?

||||

{ 218 }

Which number fits in the next spot in this sequence?:

266, 533, 1067, 2135, 4271, ...?

||||

{ 219 }

Solve this: what comes next in this sequence?:

99, 116, 133, 150, 167, ...?

||||

{ 220 }

Solve this: what comes next in this sequence?:

286, 303, 320, 337, 354, ...?

||||

{ 221 }

Which number fits in the next spot in this sequence?:

7.3, 14.6, 29.2, 58.4, 116.8, ...?

||||

{ 222 }

Which number fits in the next spot in this sequence?:

113, 227, 455, 911, 1823, ...?

||||

{ 223 }

Which number fits in the next spot in this sequence?:

163, 653, 2613, 10453, 41813, ...?

||||

{ 224 }

Which number fits in the next spot in this sequence?:

284, 853, 2560, 7681, 23044, ...?

||||

{ 225 }

Which number fits in the next spot in this sequence?:

110, 331, 994, 2983, 8950, ...?

||||

{ 226 }

Which number fits in the next spot in this sequence?:

236, 709, 2128, 6385, 19156, ...?

||||

{ 227 }

Which number fits in the next spot in this sequence?:

8.7, 34.8, 139.2, 556.8, 2227.2, ...?

||||

{ 228 }

Solve this: what comes next in this sequence?:

109, 159, 209, 259, 309, ...?

||||

{ 229 }

Which number fits in the next spot in this sequence?:

218, 873, 3493, 13973, 55893, ...?

||||

{ 230 }

Solve this: what comes next in this sequence?:

160, 206, 252, 298, 344, ...?

||||

{ 231 }

Which number fits in the next spot in this sequence?:

253, 507, 1015, 2031, 4063, ...?

||||

{ 232 }

Which number fits in the next spot in this sequence?:

167, 669, 2677, 10709, 42837, ...?

||||

{ 233 }

Which number fits in the next spot in this sequence?:

4.1, 8.2, 16.4, 32.8, 65.6, ...?

||||

{ 234 }

Solve this: what comes next in this sequence?:

251, 299, 347, 395, 443, ...?

||||

{ 235 }

Which number fits in the next spot in this sequence?:

253, 760, 2281, 6844, 20533, ...?

||||

{ 236 }

Solve this: what comes next in this sequence?:

98, 123, 148, 173, 198, ...?

||||

{ 237 }

Which number fits in the next spot in this sequence?:

4.2, 8.4, 16.8, 33.6, 67.2, ...?

||||

{ 238 }

Solve this: what comes next in this sequence?:

224, 271, 318, 365, 412, ...?

||||

{ 239 }

Solve this: what comes next in this sequence?:

141, 174, 207, 240, 273, ...?

||||

{ 240 }

Which number fits in the next spot in this sequence?:

8.1, 24.3, 72.9, 218.7, 656.1, ...?

||||

{ 241 }

Which number fits in the next spot in this sequence?:

90, 181, 363, 727, 1455, ...?

||||

{ 242 }

Solve this: what comes next in this sequence?:

285, 323, 361, 399, 437, ...?

||||

{ 243 }

Which number fits in the next spot in this sequence?:

2.1, 4.2, 8.4, 16.8, 33.6, ...?

||||

{ 244 }

Solve this: what comes next in this sequence?:

129, 176, 223, 270, 317, ...?

||||

{ 245 }

Which number fits in the next spot in this sequence?:

91, 365, 1461, 5845, 23381, ...?

||||

{ 246 }

Which number fits in the next spot in this sequence?:

97, 195, 391, 783, 1567, ...?

||||

{ 247 }

Solve this: what comes next in this sequence?:

148, 160, 172, 184, 196, ...?

||||

{ 248 }

Which number fits in the next spot in this sequence?:

8.6, 17.2, 34.4, 68.8, 137.6, ...?

||||

{ 249 }

Which number fits in the next spot in this sequence?:

144, 433, 1300, 3901, 11704, ...?

||||

{ 250 }

Which number fits in the next spot in this sequence?:

261, 1045, 4181, 16725, 66901, ...?

||||

{ 251 }

Solve this: what comes next in this sequence?:

216, 250, 284, 318, 352, ...?

||||

{ 252 }

Which number fits in the next spot in this sequence?:

7.9, 23.7, 71.1, 213.3, 639.900000000001, ...?

||||

{ 253 }

Which number fits in the next spot in this sequence?:

9.9, 19.8, 39.6, 79.2, 158.4, ...?

||||

{ 254 }

Which number fits in the next spot in this sequence?:

293, 880, 2641, 7924, 23773, ...?

||||

{ 255 }

Solve this: what comes next in this sequence:

286, 329, 372, 415, 458, ...?

||||

{ 256 }

Solve this: what comes next in this sequence?:

290, 323, 356, 389, 422, ...?

||||

{ 257 }

Solve this: what comes next in this sequence?:

217, 242, 267, 292, 317, ...?

||||

{ 258 }

Which number fits in the next spot in this sequence?:

8.7, 26.1, 78.3, 234.9, 704.7, ...?

||||

{ 259 }

Which number fits in the next spot in this sequence?:

234, 937, 3749, 14997, 59989, ...?

||||

{ 260 }

Which number fits in the next spot in this sequence?:

294, 1177, 4709, 18837, 75349, ...?

||||

{ 261 }

Which number fits in the next spot in this sequence?:

8.9, 35.6, 142.4, 569.6, 2278.4, ...?

||||

{ 262 }

Which number fits in the next spot in this sequence?:

214, 857, 3429, 13717, 54869, ...?

||||

{ 263 }

Solve this: what comes next in this sequence?:

294, 336, 378, 420, 462, ...?

||||

{ 264 }

Solve this: what comes next in this sequence?:

210, 224, 238, 252, 266, ...?

||||

{ 265 }

Which number fits in the next spot in this sequence?:

3.6, 10.8, 32.4, 97.2, 291.6, ...?

||||

{ 266 }

Solve this: what comes next in this sequence?:

160, 209, 258, 307, 356, ...?

||||

{ 267 }

Which number fits in the next spot in this sequence?:

189, 379, 759, 1519, 3039, ...?

||||

{ 268 }

Which number fits in the next spot in this sequence?:

221, 664, 1993, 5980, 17941, ...?

||||

{ 269 }

Which number fits in the next spot in this sequence?:

7.6, 22.8, 68.4, 205.2, 615.6, ...?

||||

{ 270 }

Which number fits in the next spot in this sequence?:

130, 261, 523, 1047, 2095, ...?

||||

{ 271 }

Which number fits in the next spot in this sequence?:

8.1, 16.2, 32.4, 64.8, 129.6, ...?

||||

{ 272 }

Which number fits in the next spot in this sequence?:

140, 281, 563, 1127, 2255, ...?

||||

{ 273 }

Which number fits in the next spot in this sequence?:

7.1, 14.2, 28.4, 56.8, 113.6, ...?

||||

{ 274 }

Which number fits in the next spot in this sequence?:

7.4, 29.6, 118.4, 473.6, 1894.4, ...?

||||

{ 275 }

Which number fits in the next spot in this sequence?:

2.3, 9.2, 36.8, 147.2, 588.8, ...?

||||

{ 276 }

Which number fits in the next spot in this sequence?:

125, 251, 503, 1007, 2015, ...?

||||

{ 277 }

Solve this: what comes next in this sequence?:

229, 251, 273, 295, 317, ...?

||||

{ 278 }

Which number fits in the next spot in this sequence?:

288, 1153, 4613, 18453, 73813, ...?

||||

{ 279 }

Which number fits in the next spot in this sequence?:

102, 307, 922, 2767, 8302, ...?

||||

{ 280 }

Which number fits in the next spot in this sequence?:

4.8, 9.6, 19.2, 38.4, 76.8, ...?

||||

{ 281 }

Which number fits in the next spot in this sequence?:

257, 515, 1031, 2063, 4127, ...?

||||

{ 282 }

Which number fits in the next spot in this sequence?:

177, 355, 711, 1423, 2847, ...?

||||

{ 283 }

Which number fits in the next spot in this sequence?:

8.7, 34.8, 139.2, 556.8, 2227.2, ...?

||||

{ 284 }

Which number fits in the next spot in this sequence?:

8.7, 17.4, 34.8, 69.6, 139.2, ...?

||||

{ 285 }

Which number fits in the next spot in this sequence?:

8.2, 32.8, 131.2, 524.8, 2099.2, ...?

||||

{ 286 }

Which number fits in the next spot in this sequence?:

201, 604, 1813, 5440, 16321, ...?

||||

{ 287 }

Which number fits in the next spot in this sequence?:

3.8, 11.4, 34.2, 102.6, 307.8, ...?

||||

{ 288 }

Which number fits in the next spot in this sequence?:

115, 461, 1845, 7381, 29525, ...?

||||

{ 289 }

Solve this: what comes next in this sequence?:

104, 120, 136, 152, 168, ...?

||||

{ 290 }

Solve this: what comes next in this sequence?:

108, 149, 190, 231, 272, ...?

||||

{ 291 }

Solve this: what comes next in this sequence?:

211, 224, 237, 250, 263, ...?

||||

{ 292 }

Solve this: what comes next in this sequence?:

174, 193, 212, 231, 250, ...?

||||

{ 293 }

Solve this: what comes next in this sequence?:

300, 315, 330, 345, 360, ...?

||||

{ 294 }

Which number fits in the next spot in this sequence?:

5.8, 23.2, 92.8, 371.2, 1484.8, ...?

||||

{ 295 }

Which number fits in the next spot in this sequence?:

8.1, 24.3, 72.9, 218.7, 656.1, ...?

||||

{ 296 }

Which number fits in the next spot in this sequence?:

7.9, 15.8, 31.6, 63.2, 126.4, ...?

||||

{ 297 }

Which number fits in the next spot in this sequence?:

9.6, 38.4, 153.6, 614.4, 2457.6, ...?

||||

{ 298 }

Which number fits in the next spot in this sequence?:

8.3, 24.9, 74.7, 224.1, 672.3, ...?

||||

{ 299 }

Which number fits in the next spot in this sequence?:

3.5, 7, 14, 28, 56, ...?

||||

{ 300 }

Which number fits in the next spot in this sequence?:

6.8, 20.4, 61.2, 183.6, 550.8, ...?

IIII

{ 301 }

Which number fits in the next spot in this sequence?:

98, 393, 1573, 6293, 25173, ...?

IIII

{ 302 }

Which number fits in the next spot in this sequence?:

191, 574, 1723, 5170, 15511, ...?

IIII

{ 303 }

Which number fits in the next spot in this sequence?:

186, 373, 747, 1495, 2991, ...?

IIII

{ 304 }

Which number fits in the next spot in this sequence?:

7.4, 29.6, 118.4, 473.6, 1894.4, ...?

IIII

{ 305 }

Solve this: what comes next in this sequence?:

109, 144, 179, 214, 249, ...?

||||

{ 306 }

Which number fits in the next spot in this sequence?:

7.9, 23.7, 71.1, 213.3, 639.900000000001, ...?

||||

{ 307 }

Which number fits in the next spot in this sequence?:

3.9, 15.6, 62.4, 249.6, 998.4, ...?

||||

{ 308 }

Solve this: what comes next in this sequence?:

250, 294, 338, 382, 426, ...?

||||

{ 309 }

Solve this: what comes next in this sequence?:

93, 114, 135, 156, 177, ...?

||||

{ 310 }

Solve this: what comes next in this sequence?:

285, 322, 359, 396, 433, ...?

||||

{ 311 }

Solve this: what comes next in this sequence?:

281, 305, 329, 353, 377, ...?

||||

{ 312 }

Solve this: what comes next in this sequence?:

199, 235, 271, 307, 343, ...?

||||

{ 313 }

Solve this: what comes next in this sequence?:

100, 121, 142, 163, 184, ...?

||||

{ 314 }

Solve this: what comes next in this sequence?:

230, 254, 278, 302, 326, ...?

||||

{ 315 }

Which number fits in the next spot in this sequence?:

142, 285, 571, 1143, 2287, ...?

||||

{ 316 }

Solve this: what comes next in this sequence?:

143, 189, 235, 281, 327, ...?

||||

{ 317 }

Which number fits in the next spot in this sequence?:

136, 409, 1228, 3685, 11056, ...?

||||

{ 318 }

Solve this: what comes next in this sequence?:

106, 119, 132, 145, 158, ...?

||||

{ 319 }

Which number fits in the next spot in this sequence?:

5.2, 15.6, 46.8, 140.4, 421.2, ...?

||||

{ 320 }

Which number fits in the next spot in this sequence?:

8.8, 26.4, 79.2, 237.6, 712.800000000001, ...?

||||

{ 321 }

Solve this: what comes next in this sequence?:

275, 288, 301, 314, 327, ...?

||||

{ 322 }

Solve this: what comes next in this sequence?:

203, 234, 265, 296, 327, ...?

||||

{ 323 }

Solve this: what comes next in this sequence?:

108, 152, 196, 240, 284, ...?

||||

{ 324 }

Which number fits in the next spot in this sequence?:

227, 682, 2047, 6142, 18427, ...?

||||

{ 325 }

Which number fits in the next spot in this sequence?:

253, 507, 1015, 2031, 4063, ...?

||||

{ 326 }

Solve this: what comes next in this sequence?:

242, 270, 298, 326, 354, ...?

||||

{ 327 }

Which number fits in the next spot in this sequence?:

9.1, 18.2, 36.4, 72.8, 145.6, ...?

||||

{ 328 }

Solve this: what comes next in this sequence?:

187, 209, 231, 253, 275, ...?

||||

{ 329 }

Which number fits in the next spot in this sequence?:

1.1, 3.3, 9.9, 29.7, 89.1000000000001, ...?

||||

{ 330 }

Which number fits in the next spot in this sequence?:

8.2, 24.6, 73.8, 221.4, 664.2, ...?

||||

{ 331 }

Which number fits in the next spot in this sequence?:

9.1, 27.3, 81.9, 245.7, 737.1, ...?

||||

{ 332 }

Which number fits in the next spot in this sequence?:

7, 21, 63, 189, 567, ...?

||||

{ 333 }

Which number fits in the next spot in this sequence?:

7.3, 29.2, 116.8, 467.2, 1868.8, ...?

||||

{ 334 }

Which number fits in the next spot in this sequence?:

210, 421, 843, 1687, 3375, ...?

||||

{ 335 }

Which number fits in the next spot in this sequence?:

190, 381, 763, 1527, 3055, ...?

||||

{ 336 }

Solve this: what comes next in this sequence?:

262, 282, 302, 322, 342, ...?

||||

{ 337 }

Which number fits in the next spot in this sequence?:

2.9, 8.7, 26.1, 78.3, 234.9, ...?

||||

{ 338 }

Which number fits in the next spot in this sequence?:

8.7, 17.4, 34.8, 69.6, 139.2, ...?

||||

{ 339 }

Which number fits in the next spot in this sequence?:

3.3, 13.2, 52.8, 211.2, 844.8, ...?

||||

{ 340 }

Which number fits in the next spot in this sequence?:

147, 589, 2357, 9429, 37717, ...?

IIII

{ 341 }

Solve this: what comes next in this sequence?:

150, 200, 250, 300, 350, ...?

IIII

{ 342 }

Which number fits in the next spot in this sequence?:

2.7, 10.8, 43.2, 172.8, 691.2, ...?

IIII

{ 343 }

Which number fits in the next spot in this sequence?:

3.1, 9.3, 27.9, 83.7, 251.1, ...?

IIII

{ 344 }

Which number fits in the next spot in this sequence?:

3.8, 7.6, 15.2, 30.4, 60.8, ...?

IIII

{ 345 }

Solve this: what comes next in this sequence?:

226, 240, 254, 268, 282, ...?

||||

{ 346 }

Which number fits in the next spot in this sequence?:

213, 427, 855, 1711, 3423, ...?

||||

{ 347 }

Which number fits in the next spot in this sequence?:

8.9, 26.7, 80.1, 240.3, 720.9, ...?

||||

{ 348 }

Which number fits in the next spot in this sequence?:

1.2, 3.6, 10.8, 32.4, 97.2, ...?

||||

{ 349 }

Solve this: what comes next in this sequence?:

284, 295, 306, 317, 328, ...?

||||

{ 350 }

Solve this: what comes next in this sequence?:

122, 142, 162, 182, 202, ...?

||||

{ 351 }

Which number fits in the next spot in this sequence?:

230, 461, 923, 1847, 3695, ...?

||||

{ 352 }

Which number fits in the next spot in this sequence?:

9.6, 19.2, 38.4, 76.8, 153.6, ...?

||||

{ 353 }

Which number fits in the next spot in this sequence?:

153, 307, 615, 1231, 2463, ...?

||||

{ 354 }

Solve this: what comes next in this sequence?:

170, 216, 262, 308, 354, ...?

||||

{ 355 }

Solve this: what comes next in this sequence?:

212, 262, 312, 362, 412, ...?

IIII

{ 356 }

Which number fits in the next spot in this sequence?:

3.9, 7.8, 15.6, 31.2, 62.4, ...?

IIII

{ 357 }

Which number fits in the next spot in this sequence?:

7, 14, 28, 56, 112, ...?

IIII

{ 358 }

Solve this: what comes next in this sequence?:

126, 150, 174, 198, 222, ...?

IIII

{ 359 }

Which number fits in the next spot in this sequence?:

170, 341, 683, 1367, 2735, ...?

IIII

{ 360 }

Which number fits in the next spot in this sequence?:

127, 509, 2037, 8149, 32597, ...?

||||

{ 361 }

Solve this: what comes next in this sequence?:

126, 168, 210, 252, 294, ...?

||||

{ 362 }

Which number fits in the next spot in this sequence?:

204, 613, 1840, 5521, 16564, ...?

||||

{ 363 }

Which number fits in the next spot in this sequence?:

2.5, 10, 40, 160, 640, ...?

||||

{ 364 }

Which number fits in the next spot in this sequence?:

234, 469, 939, 1879, 3759, ...?

||||

{ 365 }

Which number fits in the next spot in this sequence?:

155, 466, 1399, 4198, 12595, ...?

||||

{ 366 }

Which number fits in the next spot in this sequence?:

2.7, 10.8, 43.2, 172.8, 691.2, ...?

||||

{ 367 }

Which number fits in the next spot in this sequence?:

138, 553, 2213, 8853, 35413, ...?

||||

{ 368 }

Solve this: what comes next in this sequence?:

96, 112, 128, 144, 160, ...?

||||

{ 369 }

Solve this: what comes next in this sequence?:

117, 127, 137, 147, 157, ...?

||||

{ 370 }

Which number fits in the next spot in this sequence?:

4.8, 14.4, 43.2, 129.6, 388.8, ...?

||||

{ 371 }

Which number fits in the next spot in this sequence?:

1.9, 5.7, 17.1, 51.3, 153.9, ...?

||||

{ 372 }

Which number fits in the next spot in this sequence?:

158, 317, 635, 1271, 2543, ...?

||||

{ 373 }

Which number fits in the next spot in this sequence?:

8.2, 32.8, 131.2, 524.8, 2099.2, ...?

||||

{ 374 }

Which number fits in the next spot in this sequence?:

8.1, 24.3, 72.9, 218.7, 656.1, ...?

||||

{ 375 }

Which number fits in the next spot in this sequence?:

10, 20, 40, 80, 160, ...?

||||

{ 376 }

Which number fits in the next spot in this sequence?:

274, 1097, 4389, 17557, 70229, ...?

||||

{ 377 }

Solve this: what comes next in this sequence?:

165, 184, 203, 222, 241, ...?

||||

{ 378 }

Solve this: what comes next in this sequence?:

196, 222, 248, 274, 300, ...?

||||

{ 379 }

Which number fits in the next spot in this sequence?:

93, 280, 841, 2524, 7573, ...?

||||

{ 380 }

Which number fits in the next spot in this sequence?:

2.1, 4.2, 8.4, 16.8, 33.6, ...?

||||

{ 381 }

Solve this: what comes next in this sequence?:

291, 323, 355, 387, 419, ...?

||||

{ 382 }

Solve this: what comes next in this sequence?:

176, 207, 238, 269, 300, ...?

||||

{ 383 }

Solve this: what comes next in this sequence?:

258, 273, 288, 303, 318, ...?

||||

{ 384 }

Which number fits in the next spot in this sequence?:

6.1, 12.2, 24.4, 48.8, 97.6, ...?

||||

{ 385 }

Which number fits in the next spot in this sequence?:

6.7, 13.4, 26.8, 53.6, 107.2, ...?

||||

{ 386 }

Which number fits in the next spot in this sequence?:

138, 277, 555, 1111, 2223, ...?

||||

{ 387 }

Which number fits in the next spot in this sequence?:

169, 677, 2709, 10837, 43349, ...?

||||

{ 388 }

Which number fits in the next spot in this sequence?:

153, 613, 2453, 9813, 39253, ...?

||||

{ 389 }

Which number fits in the next spot in this sequence?:

3, 9, 27, 81, 243, ...?

||||

{ 390 }

Which number fits in the next spot in this sequence?:

3.4, 6.8, 13.6, 27.2, 54.4, ...?

||||

{ 391 }

Which number fits in the next spot in this sequence?:

225, 676, 2029, 6088, 18265, ...?

||||

{ 392 }

Which number fits in the next spot in this sequence?:

4.8, 19.2, 76.8, 307.2, 1228.8, ...?

||||

{ 393 }

Solve this: what comes next in this sequence?:

295, 344, 393, 442, 491, ...?

||||

{ 394 }

Solve this: what comes next in this sequence?:

165, 181, 197, 213, 229, ...?

||||

{ 395 }

Solve this: what comes next in this sequence?:

256, 296, 336, 376, 416, ...?

||||

{ 396 }

Which number fits in the next spot in this sequence?:

137, 549, 2197, 8789, 35157, ...?

||||

{ 397 }

Which number fits in the next spot in this sequence?:

7.1, 14.2, 28.4, 56.8, 113.6, ...?

||||

{ 398 }

Solve this: what comes next in this sequence?:

234, 276, 318, 360, 402, ...?

||||

{ 399 }

Solve this: what comes next in this sequence?:

251, 280, 309, 338, 367, ...?

||||

{ 400 }

Solve this: what comes next in this sequence?:

244, 286, 328, 370, 412, ...?

||||

{ 401 }

Which number fits in the next spot in this sequence?:

5, 15, 45, 135, 405, ...?

||||

{ 402 }

Which number fits in the next spot in this sequence?:

119, 358, 1075, 3226, 9679, ...?

||||

{ 403 }

Which number fits in the next spot in this sequence?:

9.5, 28.5, 85.5, 256.5, 769.5, ...?

||||

{ 404 }

Which number fits in the next spot in this sequence?:

290, 581, 1163, 2327, 4655, ...?

||||

{ 405 }

Which number fits in the next spot in this sequence?:

198, 595, 1786, 5359, 16078, ...?

||||

{ 406 }

Which number fits in the next spot in this sequence?:

232, 697, 2092, 6277, 18832, ...?

||||

{ 407 }

Which number fits in the next spot in this sequence?:

8.5, 34, 136, 544, 2176, ...?

||||

{ 408 }

Which number fits in the next spot in this sequence?:

5.2, 15.6, 46.8, 140.4, 421.2, ...?

||||

{ 409 }

Which number fits in the next spot in this sequence?:

288, 1153, 4613, 18453, 73813, ...?

||||

{ 410 }

Which number fits in the next spot in this sequence?:

6.5, 19.5, 58.5, 175.5, 526.5, ...?

||||

{ 411 }

Solve this: what comes next in this sequence?:

94, 117, 140, 163, 186, ...?

||||

{ 412 }

Which number fits in the next spot in this sequence?:

231, 463, 927, 1855, 3711, ...?

||||

{ 413 }

Which number fits in the next spot in this sequence?:

9.3, 37.2, 148.8, 595.2, 2380.8, ...?

||||

{ 414 }

Which number fits in the next spot in this sequence?:

238, 953, 3813, 15253, 61013, ...?

||||

{ 415 }

Which number fits in the next spot in this sequence?:

1.2, 2.4, 4.8, 9.6, 19.2, ...?

||||

{ 416 }

Which number fits in the next spot in this sequence?:

173, 520, 1561, 4684, 14053, ...?

||||

{ 417 }

Solve this: what comes next in this sequence?:

210, 231, 252, 273, 294, ...?

||||

{ 418 }

Which number fits in the next spot in this sequence?:

242, 727, 2182, 6547, 19642, ...?

||||

{ 419 }

Solve this: what comes next in this sequence?:

124, 147, 170, 193, 216, ...?

||||

{ 420 }

Which number fits in the next spot in this sequence?:

118, 473, 1893, 7573, 30293, ...?

||||

{ 21 }

Which number fits in the next spot in this sequence?:

160, 641, 2565, 10261, 41045, ...?

||||

By observation and trial and error, we find that each term in this

sequence is obtained by multiplying the previous term by 4 and adding 1.

Therefore, the sixth term = (4 x 41045) + 1 = 164181.

|||||||||

{ 22 }

Which number fits in the next spot in this sequence?:

93, 280, 841, 2524, 7573, ...?

||||

By observation and trial and error, we find that each term in this

sequence is obtained by multiplying the previous term by 3 and adding 1.

Therefore, the sixth term = (3 x 7573) + 1 = 22720.

||||||||

{ 23 }

Which number fits in the next spot in this sequence?:

145, 291, 583, 1167, 2335, ...?

||||

By observation and trial and error, we find that each term in this sequence is obtained by multiplying the previous term by 2 and adding 1. Therefore, the sixth term = (2 x 2335) + 1 = 4671.

|||||||||

{ 24 }

Which number fits in the next spot in this sequence?:

189, 568, 1705, 5116, 15349, ...?

||||

By observation and trial and error, we find that each term in this sequence is obtained by multiplying the previous term by 3 and adding 1. Therefore, the sixth term = (3 x 15349) + 1 = 46048.

|||||||||

{ 25 }

Solve this: what comes next in this sequence?:

144, 180, 216, 252, 288, ...?

||||

By taking the differences between adjacent terms, we can see that this is an Arithmetic Progression, with common difference = 36. Therefore, the sixth term can be calculated by adding the common difference to the fifth

term. It is equal to 288 + 36 = 324.

||||||||

{ 26 }

Which number fits in the next spot in this sequence?:

9.6, 28.8, 86.4, 259.2, 777.6, ...?

||||

By taking the ratios of adjacent terms, we can see that this is a

Geometric Progression, with common ratio = 3. Therefore, the sixth term

can be calculated by multiplying the fifth term with the common ratio. It is

equal to 777.6 x 3 = 2332.8.

||||||||

{ 27 }

Solve this: what comes next in this sequence?:

150, 176, 202, 228, 254, ...?

||||

By taking the differences between adjacent terms, we can see that this is

an Arithmetic Progression, with common difference = 26. Therefore, the

sixth term can be calculated by adding the common difference to the fifth

term. It is equal to 254 + 26 = 280.

||||||||

{ 28 }

Which number fits in the next spot in this sequence?:

3.2, 9.6, 28.8, 86.4, 259.2, ...?

||||

By taking the ratios of adjacent terms, we can see that this is a

Geometric Progression, with common ratio = 3. Therefore, the sixth term

can be calculated by multiplying the fifth term with the common ratio. It is

equal to 259.2 x 3 = 777.6.

|||||||||

{ 29 }

Which number fits in the next spot in this sequence?:

1.1, 4.4, 17.6, 70.4, 281.6, ...?

||||

By taking the ratios of adjacent terms, we can see that this is a

Geometric Progression, with common ratio = 4. Therefore, the sixth term

can be calculated by multiplying the fifth term with the common ratio. It is

equal to 281.6 x 4 = 1126.4.

|||||||||

{ 30 }

Which number fits in the next spot in this sequence?:

7.4, 29.6, 118.4, 473.6, 1894.4, ...?

||||

By taking the ratios of adjacent terms, we can see that this is a

Geometric Progression, with common ratio = 4. Therefore, the sixth term

can be calculated by multiplying the fifth term with the common ratio. It is

equal to 1894.4 x 4 = 7577.6.

|||||||||

{ 31 }

Which number fits in the next spot in this sequence?:

238, 715, 2146, 6439, 19318, ...?

||||

By observation and trial and error, we find that each term in this

sequence is obtained by multiplying the previous term by 3 and adding 1.

Therefore, the sixth term = (3 x 19318) + 1 = 57955.

|||||||||

{ 32 }

Which number fits in the next spot in this sequence?:

2, 8, 32, 128, 512, ...?

||||

By taking the ratios of adjacent terms, we can see that this is a

Geometric Progression, with common ratio = 4. Therefore, the sixth term

can be calculated by multiplying the fifth term with the common ratio. It is

equal to 512 x 4 = 2048.

||||||||

{ 33 }

Which number fits in the next spot in this sequence?:

6.9, 20.7, 62.1, 186.3, 558.9, ...?

||||

By taking the ratios of adjacent terms, we can see that this is a

Geometric Progression, with common ratio = 3. Therefore, the sixth term

can be calculated by multiplying the fifth term with the common ratio. It is

equal to 558.9 x 3 = 1676.7.

||||||||

{ 34 }

Solve this: what comes next in this sequence?:

107, 133, 159, 185, 211, ...?

||||

By taking the differences between adjacent terms, we can see that this is

an Arithmetic Progression, with common difference = 26. Therefore, the

sixth term can be calculated by adding the common difference to the fifth

term. It is equal to 211 + 26 = 237.

||||||||

{ 35 }

Which number fits in the next spot in this sequence?:

7.7, 30.8, 123.2, 492.8, 1971.2, ...?

||||

By taking the ratios of adjacent terms, we can see that this is a

Geometric Progression, with common ratio = 4. Therefore, the sixth term

can be calculated by multiplying the fifth term with the common ratio. It is

equal to 1971.2 x 4 = 7884.8.

|||||||||

{ 36 }

Solve this: what comes next in this sequence?:

298, 320, 342, 364, 386, ...?

||||

By taking the differences between adjacent terms, we can see that this is

an Arithmetic Progression, with common difference = 22. Therefore, the

sixth term can be calculated by adding the common difference to the fifth

term. It is equal to 386 + 22 = 408.

|||||||||

{ 37 }

Which number fits in the next spot in this sequence?:

8.8, 17.6, 35.2, 70.4, 140.8, ...?

||||

By taking the ratios of adjacent terms, we can see that this is a

Geometric Progression, with common ratio = 2. Therefore, the sixth term

can be calculated by multiplying the fifth term with the common ratio. It is

equal to 140.8 x 2 = 281.6.

|||||||||

{ 38 }

Which number fits in the next spot in this sequence?:

7.8, 23.4, 70.2, 210.6, 631.8, ...?

||||

By taking the ratios of adjacent terms, we can see that this is a

Geometric Progression, with common ratio = 3. Therefore, the sixth term

can be calculated by multiplying the fifth term with the common ratio. It is

equal to 631.8 x 3 = 1895.4.

|||||||||

{ 39 }

Which number fits in the next spot in this sequence?:

220, 661, 1984, 5953, 17860, ...?

||||

By observation and trial and error, we find that each term in this

sequence is obtained by multiplying the previous term by 3 and adding 1.

Therefore, the sixth term = (3 x 17860) + 1 = 53581.

IIIIIIII

{ 40 }

Which number fits in the next spot in this sequence?:

164, 329, 659, 1319, 2639, ...?

IIII

By observation and trial and error, we find that each term in this

sequence is obtained by multiplying the previous term by 2 and adding 1.

Therefore, the sixth term = (2 x 2639) + 1 = 5279.

IIIIIIII

{ 41 }

Solve this: what comes next in this sequence?:

268, 315, 362, 409, 456, ...?

IIII

By taking the differences between adjacent terms, we can see that this is

an Arithmetic Progression, with common difference = 47. Therefore, the

sixth term can be calculated by adding the common difference to the fifth

term. It is equal to 456 + 47 = 503.

IIIIIIII

{ 42 }

Which number fits in the next spot in this sequence?:

124, 373, 1120, 3361, 10084, ...?

||||

By observation and trial and error, we find that each term in this

sequence is obtained by multiplying the previous term by 3 and adding 1.

Therefore, the sixth term = (3 x 10084) + 1 = 30253.

|||||||||

{ 43 }

Which number fits in the next spot in this sequence?:

4.1, 12.3, 36.9, 110.7, 332.1, ...?

||||

By taking the ratios of adjacent terms, we can see that this is a

Geometric Progression, with common ratio = 3. Therefore, the sixth term

can be calculated by multiplying the fifth term with the common ratio. It is

equal to 332.1 x 3 = 996.3.

|||||||||

{ 44 }

Solve this: what comes next in this sequence?:

258, 291, 324, 357, 390, ...?

||||

By taking the differences between adjacent terms, we can see that this is an Arithmetic Progression, with common difference = 33. Therefore, the sixth term can be calculated by adding the common difference to the fifth term. It is equal to 390 + 33 = 423.

||||||||

{ 45 }

Which number fits in the next spot in this sequence?:

9.1, 18.2, 36.4, 72.8, 145.6, ...?

||||

By taking the ratios of adjacent terms, we can see that this is a Geometric Progression, with common ratio = 2. Therefore, the sixth term can be calculated by multiplying the fifth term with the common ratio. It is equal to 145.6 x 2 = 291.2.

||||||||

{ 46 }

Which number fits in the next spot in this sequence?:

99, 199, 399, 799, 1599, ...?

||||

By observation and trial and error, we find that each term in this sequence is obtained by multiplying the previous term by 2 and adding 1. Therefore, the sixth term = (2 x 1599) + 1 = 3199.

IIIIIII

{ 47 }

Solve this: what comes next in this sequence?:

153, 203, 253, 303, 353, ...?

IIII

By taking the differences between adjacent terms, we can see that this is an Arithmetic Progression, with common difference = 50. Therefore, the sixth term can be calculated by adding the common difference to the fifth term. It is equal to 353 + 50 = 403.

IIIIIII

{ 48 }

Which number fits in the next spot in this sequence?:

5, 15, 45, 135, 405, ...?

IIII

By taking the ratios of adjacent terms, we can see that this is a Geometric Progression, with common ratio = 3. Therefore, the sixth term can be calculated by multiplying the fifth term with the common ratio. It is equal to 405 x 3 = 1215.

IIIIIII

{ 49 }

Which number fits in the next spot in this sequence?:

6.7, 13.4, 26.8, 53.6, 107.2, ...?

||||

By taking the ratios of adjacent terms, we can see that this is a

Geometric Progression, with common ratio = 2. Therefore, the sixth term

can be calculated by multiplying the fifth term with the common ratio. It is

equal to 107.2 x 2 = 214.4.

|||||||||

{ 50 }

Which number fits in the next spot in this sequence?:

8.2, 24.6, 73.8, 221.4, 664.2, ...?

||||

By taking the ratios of adjacent terms, we can see that this is a

Geometric Progression, with common ratio = 3. Therefore, the sixth term

can be calculated by multiplying the fifth term with the common ratio. It is

equal to 664.2 x 3 = 1992.6.

|||||||||

{ 51 }

Which number fits in the next spot in this sequence?:

5.6, 22.4, 89.6, 358.4, 1433.6, ...?

||||

By taking the ratios of adjacent terms, we can see that this is a

Geometric Progression, with common ratio = 4. Therefore, the sixth term

can be calculated by multiplying the fifth term with the common ratio. It is

equal to 1433.6 x 4 = 5734.4.

|||||||||

{ 52 }

Solve this: what comes next in this sequence?:

238, 277, 316, 355, 394, ...?

||||

By taking the differences between adjacent terms, we can see that this is

an Arithmetic Progression, with common difference = 39. Therefore, the

sixth term can be calculated by adding the common difference to the fifth

term. It is equal to 394 + 39 = 433.

|||||||||

{ 53 }

Solve this: what comes next in this sequence?:

153, 191, 229, 267, 305, ...?

||||

By taking the differences between adjacent terms, we can see that this is

an Arithmetic Progression, with common difference = 38. Therefore, the

sixth term can be calculated by adding the common difference to the fifth

term. It is equal to 305 + 38 = 343.

||||||||

{ 54 }

Solve this: what comes next in this sequence?:

214, 224, 234, 244, 254, ...?

||||

By taking the differences between adjacent terms, we can see that this is

an Arithmetic Progression, with common difference = 10. Therefore, the

sixth term can be calculated by adding the common difference to the fifth

term. It is equal to 254 + 10 = 264.

||||||||

{ 55 }

Which number fits in the next spot in this sequence?:

1.2, 4.8, 19.2, 76.8, 307.2, ...?

||||

By taking the ratios of adjacent terms, we can see that this is a

Geometric Progression, with common ratio = 4. Therefore, the sixth term

can be calculated by multiplying the fifth term with the common ratio. It is

equal to 307.2 x 4 = 1228.8.

||||||||

{ 56 }

Which number fits in the next spot in this sequence?:

4.6, 9.2, 18.4, 36.8, 73.6, ...?

||||

By taking the ratios of adjacent terms, we can see that this is a

Geometric Progression, with common ratio = 2. Therefore, the sixth term

can be calculated by multiplying the fifth term with the common ratio. It is

equal to 73.6 x 2 = 147.2.

|||||||||

{ 57 }

Solve this: what comes next in this sequence?:

171, 194, 217, 240, 263, ...?

||||

By taking the differences between adjacent terms, we can see that this is

an Arithmetic Progression, with common difference = 23. Therefore, the

sixth term can be calculated by adding the common difference to the fifth

term. It is equal to 263 + 23 = 286.

|||||||||

{ 58 }

Which number fits in the next spot in this sequence?:

155, 466, 1399, 4198, 12595, ...?

||||

By observation and trial and error, we find that each term in this sequence is obtained by multiplying the previous term by 3 and adding 1. Therefore, the sixth term = (3 x 12595) + 1 = 37786.

||||||||

{ 59 }

Which number fits in the next spot in this sequence?:

6.1, 18.3, 54.9, 164.7, 494.1, ...?

||||

By taking the ratios of adjacent terms, we can see that this is a Geometric Progression, with common ratio = 3. Therefore, the sixth term can be calculated by multiplying the fifth term with the common ratio. It is equal to 494.1 x 3 = 1482.3.

||||||||

{ 60 }

Which number fits in the next spot in this sequence?:

257, 1029, 4117, 16469, 65877, ...?

||||

By observation and trial and error, we find that each term in this sequence is obtained by multiplying the previous term by 4 and adding 1. Therefore, the sixth term = (4 x 65877) + 1 = 263509.

||||||||

{ 61 }

Which number fits in the next spot in this sequence?:

9.7, 19.4, 38.8, 77.6, 155.2, ...?

||||

By taking the ratios of adjacent terms, we can see that this is a

Geometric Progression, with common ratio = 2. Therefore, the sixth term

can be calculated by multiplying the fifth term with the common ratio. It is

equal to 155.2 x 2 = 310.4.

|||||||||

{ 62 }

Which number fits in the next spot in this sequence?:

124, 497, 1989, 7957, 31829, ...?

||||

By observation and trial and error, we find that each term in this

sequence is obtained by multiplying the previous term by 4 and adding 1.

Therefore, the sixth term = (4 x 31829) + 1 = 127317.

|||||||||

{ 63 }

Solve this: what comes next in this sequence?:

191, 210, 229, 248, 267, ...?

||||

By taking the differences between adjacent terms, we can see that this is an Arithmetic Progression, with common difference = 19. Therefore, the sixth term can be calculated by adding the common difference to the fifth term. It is equal to 267 + 19 = 286.

|||||||||

{ 64 }

Solve this: what comes next in this sequence?:

192, 226, 260, 294, 328, ...?

||||

By taking the differences between adjacent terms, we can see that this is an Arithmetic Progression, with common difference = 34. Therefore, the sixth term can be calculated by adding the common difference to the fifth term. It is equal to 328 + 34 = 362.

|||||||||

{ 65 }

Which number fits in the next spot in this sequence?:

3.7, 7.4, 14.8, 29.6, 59.2, ...?

||||

By taking the ratios of adjacent terms, we can see that this is a

Geometric Progression, with common ratio = 2. Therefore, the sixth term

can be calculated by multiplying the fifth term with the common ratio. It is

equal to 59.2 x 2 = 118.4.

|||||||||

{ 66 }

Which number fits in the next spot in this sequence?:

110, 221, 443, 887, 1775, ...?

||||

By observation and trial and error, we find that each term in this

sequence is obtained by multiplying the previous term by 2 and adding 1.

Therefore, the sixth term = (2 x 1775) + 1 = 3551.

|||||||||

{ 67 }

Which number fits in the next spot in this sequence?:

5.2, 15.6, 46.8, 140.4, 421.2, ...?

||||

By taking the ratios of adjacent terms, we can see that this is a

Geometric Progression, with common ratio = 3. Therefore, the sixth term

can be calculated by multiplying the fifth term with the common ratio. It is

equal to 421.2 x 3 = 1263.6.

||||||||

{ 68 }

Solve this: what comes next in this sequence?:

294, 339, 384, 429, 474, ...?

||||

By taking the differences between adjacent terms, we can see that this is an Arithmetic Progression, with common difference = 45. Therefore, the sixth term can be calculated by adding the common difference to the fifth term. It is equal to 474 + 45 = 519.

||||||||

{ 69 }

Which number fits in the next spot in this sequence?:

6.6, 19.8, 59.4, 178.2, 534.6, ...?

||||

By taking the ratios of adjacent terms, we can see that this is a Geometric Progression, with common ratio = 3. Therefore, the sixth term can be calculated by multiplying the fifth term with the common ratio. It is equal to 534.6 x 3 = 1603.8.

||||||||

{ 70 }

Which number fits in the next spot in this sequence?:

131, 394, 1183, 3550, 10651, ...?

||||

By observation and trial and error, we find that each term in this

sequence is obtained by multiplying the previous term by 3 and adding 1.

Therefore, the sixth term = (3 x 10651) + 1 = 31954.

|||||||||

{ 71 }

Which number fits in the next spot in this sequence?:

8.3, 33.2, 132.8, 531.2, 2124.8, ...?

||||

By taking the ratios of adjacent terms, we can see that this is a

Geometric Progression, with common ratio = 4. Therefore, the sixth term

can be calculated by multiplying the fifth term with the common ratio. It is

equal to 2124.8 x 4 = 8499.2.

|||||||||

{ 72 }

Solve this: what comes next in this sequence?:

230, 262, 294, 326, 358, ...?

||||

By taking the differences between adjacent terms, we can see that this is an Arithmetic Progression, with common difference = 32. Therefore, the sixth term can be calculated by adding the common difference to the fifth term. It is equal to 358 + 32 = 390.

||||||||

{ 73 }

Which number fits in the next spot in this sequence?:

223, 670, 2011, 6034, 18103, ...?

||||

By observation and trial and error, we find that each term in this sequence is obtained by multiplying the previous term by 3 and adding 1. Therefore, the sixth term = (3 x 18103) + 1 = 54310.

|||||||||

{ 74 }

Which number fits in the next spot in this sequence?:

1.6, 6.4, 25.6, 102.4, 409.6, ...?

||||

By taking the ratios of adjacent terms, we can see that this is a Geometric Progression, with common ratio = 4. Therefore, the sixth term can be calculated by multiplying the fifth term with the common ratio. It is equal to 409.6 x 4 = 1638.4.

|||||||||

{ 75 }

Solve this: what comes next in this sequence?:

173, 220, 267, 314, 361, ...?

||||

By taking the differences between adjacent terms, we can see that this is

an Arithmetic Progression, with common difference = 47. Therefore, the

sixth term can be calculated by adding the common difference to the fifth

term. It is equal to 361 + 47 = 408.

|||||||||

{ 76 }

Which number fits in the next spot in this sequence?:

278, 835, 2506, 7519, 22558, ...?

||||

By observation and trial and error, we find that each term in this

sequence is obtained by multiplying the previous term by 3 and adding 1.

Therefore, the sixth term = (3 x 22558) + 1 = 67675.

|||||||||

{ 77 }

Which number fits in the next spot in this sequence?:

157, 472, 1417, 4252, 12757, ...?

||||

By observation and trial and error, we find that each term in this sequence is obtained by multiplying the previous term by 3 and adding 1. Therefore, the sixth term = (3 x 12757) + 1 = 38272.

|||||||||

{ 78 }

Solve this: what comes next in this sequence?:

280, 297, 314, 331, 348, ...?

||||

By taking the differences between adjacent terms, we can see that this is an Arithmetic Progression, with common difference = 17. Therefore, the sixth term can be calculated by adding the common difference to the fifth term. It is equal to 348 + 17 = 365.

|||||||||

{ 79 }

Which number fits in the next spot in this sequence?:

223, 447, 895, 1791, 3583, ...?

||||

By observation and trial and error, we find that each term in this sequence is obtained by multiplying the previous term by 2 and adding 1. Therefore, the sixth term = (2 x 3583) + 1 = 7167.

||||||||

{ 80 }

Which number fits in the next spot in this sequence?:

5, 20, 80, 320, 1280, ...?

||||

By taking the ratios of adjacent terms, we can see that this is a Geometric Progression, with common ratio = 4. Therefore, the sixth term can be calculated by multiplying the fifth term with the common ratio. It is equal to 1280 x 4 = 5120.

|||||||||

{ 81 }

Solve this: what comes next in this sequence?:

100, 141, 182, 223, 264, ...?

||||

By taking the differences between adjacent terms, we can see that this is an Arithmetic Progression, with common difference = 41. Therefore, the sixth term can be calculated by adding the common difference to the fifth term. It is equal to 264 + 41 = 305.

||||||||

{ 82 }

Solve this: what comes next in this sequence?:

185, 233, 281, 329, 377, ...?

||||

By taking the differences between adjacent terms, we can see that this is

an Arithmetic Progression, with common difference = 48. Therefore, the

sixth term can be calculated by adding the common difference to the fifth

term. It is equal to 377 + 48 = 425.

||||||||

{ 83 }

Which number fits in the next spot in this sequence?:

4.9, 14.7, 44.1, 132.3, 396.9, ...?

||||

By taking the ratios of adjacent terms, we can see that this is a

Geometric Progression, with common ratio = 3. Therefore, the sixth term

can be calculated by multiplying the fifth term with the common ratio. It is

equal to 396.9 x 3 = 1190.7.

||||||||

{ 84 }

Which number fits in the next spot in this sequence?:

266, 1065, 4261, 17045, 68181, ...?

||||

By observation and trial and error, we find that each term in this

sequence is obtained by multiplying the previous term by 4 and adding 1.

Therefore, the sixth term = (4 x 68181) + 1 = 272725.

|||||||||

{ 85 }

Solve this: what comes next in this sequence?:

100, 137, 174, 211, 248, ...?

||||

By taking the differences between adjacent terms, we can see that this is

an Arithmetic Progression, with common difference = 37. Therefore, the

sixth term can be calculated by adding the common difference to the fifth

term. It is equal to 248 + 37 = 285.

|||||||||

{ 86 }

Solve this: what comes next in this sequence?:

287, 317, 347, 377, 407, ...?

||||

By taking the differences between adjacent terms, we can see that this is an Arithmetic Progression, with common difference = 30. Therefore, the sixth term can be calculated by adding the common difference to the fifth term. It is equal to 407 + 30 = 437.

IIIIIIII

{ 87 }

Which number fits in the next spot in this sequence?:

4.6, 13.8, 41.4, 124.2, 372.6, ...?

IIII

By taking the ratios of adjacent terms, we can see that this is a Geometric Progression, with common ratio = 3. Therefore, the sixth term can be calculated by multiplying the fifth term with the common ratio. It is equal to 372.6 x 3 = 1117.8.

IIIIIIII

{ 88 }

Solve this: what comes next in this sequence?:

132, 162, 192, 222, 252, ...?

IIII

By taking the differences between adjacent terms, we can see that this is an Arithmetic Progression, with common difference = 30. Therefore, the sixth term can be calculated by adding the common difference to the fifth

term. It is equal to 252 + 30 = 282.

||||||||

{ 89 }

Which number fits in the next spot in this sequence?:

276, 553, 1107, 2215, 4431, ...?

||||

By observation and trial and error, we find that each term in this

sequence is obtained by multiplying the previous term by 2 and adding 1.

Therefore, the sixth term = (2 x 4431) + 1 = 8863.

||||||||

{ 90 }

Which number fits in the next spot in this sequence?:

5.5, 16.5, 49.5, 148.5, 445.5, ...?

||||

By taking the ratios of adjacent terms, we can see that this is a

Geometric Progression, with common ratio = 3. Therefore, the sixth term

can be calculated by multiplying the fifth term with the common ratio. It is

equal to 445.5 x 3 = 1336.5.

||||||||

{ 91 }

Solve this: what comes next in this sequence?:

296, 318, 340, 362, 384, ...?

||||

By taking the differences between adjacent terms, we can see that this is

an Arithmetic Progression, with common difference = 22. Therefore, the

sixth term can be calculated by adding the common difference to the fifth

term. It is equal to 384 + 22 = 406.

||||||||

{ 92 }

Solve this: what comes next in this sequence?:

188, 219, 250, 281, 312, ...?

||||

By taking the differences between adjacent terms, we can see that this is

an Arithmetic Progression, with common difference = 31. Therefore, the

sixth term can be calculated by adding the common difference to the fifth

term. It is equal to 312 + 31 = 343.

||||||||

{ 93 }

Which number fits in the next spot in this sequence?:

7.3, 14.6, 29.2, 58.4, 116.8, ...?

||||

By taking the ratios of adjacent terms, we can see that this is a

Geometric Progression, with common ratio = 2. Therefore, the sixth term

can be calculated by multiplying the fifth term with the common ratio. It is

equal to 116.8 x 2 = 233.6.

|||||||||

{ 94 }

Which number fits in the next spot in this sequence?:

1.6, 3.2, 6.4, 12.8, 25.6, ...?

||||

By taking the ratios of adjacent terms, we can see that this is a

Geometric Progression, with common ratio = 2. Therefore, the sixth term

can be calculated by multiplying the fifth term with the common ratio. It is

equal to 25.6 x 2 = 51.2.

|||||||||

{ 95 }

Solve this: what comes next in this sequence?:

280, 329, 378, 427, 476, ...?

||||

By taking the differences between adjacent terms, we can see that this is

an Arithmetic Progression, with common difference = 49. Therefore, the

sixth term can be calculated by adding the common difference to the fifth

term. It is equal to 476 + 49 = 525.

||||||||

{ 96 }

Which number fits in the next spot in this sequence?:

1.5, 6, 24, 96, 384, ...?

||||

By taking the ratios of adjacent terms, we can see that this is a

Geometric Progression, with common ratio = 4. Therefore, the sixth term

can be calculated by multiplying the fifth term with the common ratio. It is

equal to 384 x 4 = 1536.

||||||||

{ 97 }

Solve this: what comes next in this sequence?:

211, 259, 307, 355, 403, ...?

||||

By taking the differences between adjacent terms, we can see that this is

an Arithmetic Progression, with common difference = 48. Therefore, the

sixth term can be calculated by adding the common difference to the fifth

term. It is equal to 403 + 48 = 451.

||||||||

{ 98 }

Which number fits in the next spot in this sequence?:

162, 487, 1462, 4387, 13162, ...?

||||

By observation and trial and error, we find that each term in this

sequence is obtained by multiplying the previous term by 3 and adding 1.

Therefore, the sixth term = (3 x 13162) + 1 = 39487.

|||||||||

{ 99 }

Which number fits in the next spot in this sequence?:

205, 616, 1849, 5548, 16645, ...?

||||

By observation and trial and error, we find that each term in this

sequence is obtained by multiplying the previous term by 3 and adding 1.

Therefore, the sixth term = (3 x 16645) + 1 = 49936.

|||||||||

{ 100 }

Solve this: what comes next in this sequence?:

250, 293, 336, 379, 422, ...?

||||

By taking the differences between adjacent terms, we can see that this is

an Arithmetic Progression, with common difference = 43. Therefore, the sixth term can be calculated by adding the common difference to the fifth term. It is equal to 422 + 43 = 465.

||||||||

{ 101 }

Which number fits in the next spot in this sequence?:

6.7, 26.8, 107.2, 428.8, 1715.2, ...?

||||

By taking the ratios of adjacent terms, we can see that this is a Geometric Progression, with common ratio = 4. Therefore, the sixth term can be calculated by multiplying the fifth term with the common ratio. It is equal to 1715.2 x 4 = 6860.8.

||||||||

{ 102 }

Which number fits in the next spot in this sequence?:

249, 748, 2245, 6736, 20209, ...?

||||

By observation and trial and error, we find that each term in this sequence is obtained by multiplying the previous term by 3 and adding 1. Therefore, the sixth term = (3 x 20209) + 1 = 60628.

||||||||

{ 103 }

Which number fits in the next spot in this sequence?:

7, 21, 63, 189, 567, ...?

||||

By taking the ratios of adjacent terms, we can see that this is a

Geometric Progression, with common ratio = 3. Therefore, the sixth term

can be calculated by multiplying the fifth term with the common ratio. It is

equal to 567 x 3 = 1701.

|||||||||

{ 104 }

Solve this: what comes next in this sequence?:

165, 211, 257, 303, 349, ...?

||||

By taking the differences between adjacent terms, we can see that this is

an Arithmetic Progression, with common difference = 46. Therefore, the

sixth term can be calculated by adding the common difference to the fifth

term. It is equal to 349 + 46 = 395.

|||||||||

{ 105 }

Solve this: what comes next in this sequence?:

279, 292, 305, 318, 331, ...?

||||

By taking the differences between adjacent terms, we can see that this is an Arithmetic Progression, with common difference = 13. Therefore, the sixth term can be calculated by adding the common difference to the fifth term. It is equal to 331 + 13 = 344.

||||||||

{ 106 }

Which number fits in the next spot in this sequence?:

193, 773, 3093, 12373, 49493, ...?

||||

By observation and trial and error, we find that each term in this sequence is obtained by multiplying the previous term by 4 and adding 1. Therefore, the sixth term = (4 x 49493) + 1 = 197973.

||||||||

{ 107 }

Which number fits in the next spot in this sequence?:

159, 319, 639, 1279, 2559, ...?

||||

By observation and trial and error, we find that each term in this sequence is obtained by multiplying the previous term by 2 and adding 1. Therefore, the sixth term = (2 x 2559) + 1 = 5119.

|||||||

{ 108 }

Which number fits in the next spot in this sequence?:

5.3, 15.9, 47.7, 143.1, 429.3, ...?

||||

By taking the ratios of adjacent terms, we can see that this is a

Geometric Progression, with common ratio = 3. Therefore, the sixth term

can be calculated by multiplying the fifth term with the common ratio. It is

equal to 429.3 x 3 = 1287.9.

|||||||||

{ 109 }

Solve this: what comes next in this sequence?:

239, 265, 291, 317, 343, ...?

||||

By taking the differences between adjacent terms, we can see that this is

an Arithmetic Progression, with common difference = 26. Therefore, the

sixth term can be calculated by adding the common difference to the fifth

term. It is equal to 343 + 26 = 369.

|||||||||

{ 110 }

Which number fits in the next spot in this sequence?:

6.2, 12.4, 24.8, 49.6, 99.2, ...?

||||

By taking the ratios of adjacent terms, we can see that this is a

Geometric Progression, with common ratio = 2. Therefore, the sixth term

can be calculated by multiplying the fifth term with the common ratio. It is

equal to 99.2 x 2 = 198.4.

|||||||||

{ 111 }

Which number fits in the next spot in this sequence?:

7, 21, 63, 189, 567, ...?

||||

By taking the ratios of adjacent terms, we can see that this is a

Geometric Progression, with common ratio = 3. Therefore, the sixth term

can be calculated by multiplying the fifth term with the common ratio. It is

equal to 567 x 3 = 1701.

|||||||||

{ 112 }

Which number fits in the next spot in this sequence?:

8.8, 17.6, 35.2, 70.4, 140.8, ...?

||||

By taking the ratios of adjacent terms, we can see that this is a

Geometric Progression, with common ratio = 2. Therefore, the sixth term

can be calculated by multiplying the fifth term with the common ratio. It is

equal to 140.8 x 2 = 281.6.

|||||||||

{ 113 }

Which number fits in the next spot in this sequence?:

216, 649, 1948, 5845, 17536, ...?

||||

By observation and trial and error, we find that each term in this

sequence is obtained by multiplying the previous term by 3 and adding 1.

Therefore, the sixth term = (3 x 17536) + 1 = 52609.

|||||||||

{ 114 }

Which number fits in the next spot in this sequence?:

5.8, 23.2, 92.8, 371.2, 1484.8, ...?

||||

By taking the ratios of adjacent terms, we can see that this is a

Geometric Progression, with common ratio = 4. Therefore, the sixth term

can be calculated by multiplying the fifth term with the common ratio. It is

equal to 1484.8 x 4 = 5939.2.

|||||||||

{ 115 }

Which number fits in the next spot in this sequence?:

180, 361, 723, 1447, 2895, ...?

||||

By observation and trial and error, we find that each term in this

sequence is obtained by multiplying the previous term by 2 and adding 1.

Therefore, the sixth term = (2 x 2895) + 1 = 5791.

|||||||||

{ 116 }

Which number fits in the next spot in this sequence?:

9.6, 19.2, 38.4, 76.8, 153.6, ...?

||||

By taking the ratios of adjacent terms, we can see that this is a

Geometric Progression, with common ratio = 2. Therefore, the sixth term

can be calculated by multiplying the fifth term with the common ratio. It is

equal to 153.6 x 2 = 307.2.

|||||||||

{ 117 }

Which number fits in the next spot in this sequence?:

4.3, 8.6, 17.2, 34.4, 68.8, ...?

||||

By taking the ratios of adjacent terms, we can see that this is a

Geometric Progression, with common ratio = 2. Therefore, the sixth term

can be calculated by multiplying the fifth term with the common ratio. It is

equal to 68.8 x 2 = 137.6.

|||||||||

{ 118 }

Solve this: what comes next in this sequence?:

254, 266, 278, 290, 302, ...?

||||

By taking the differences between adjacent terms, we can see that this is

an Arithmetic Progression, with common difference = 12. Therefore, the

sixth term can be calculated by adding the common difference to the fifth

term. It is equal to 302 + 12 = 314.

|||||||||

{ 119 }

Which number fits in the next spot in this sequence?:

5.8, 23.2, 92.8, 371.2, 1484.8, ...?

||||

By taking the ratios of adjacent terms, we can see that this is a

Geometric Progression, with common ratio = 4. Therefore, the sixth term

can be calculated by multiplying the fifth term with the common ratio. It is

equal to 1484.8 x 4 = 5939.2.

|||||||||

{ 120 }

Which number fits in the next spot in this sequence?:

298, 597, 1195, 2391, 4783, ...?

||||

By observation and trial and error, we find that each term in this

sequence is obtained by multiplying the previous term by 2 and adding 1.

Therefore, the sixth term = (2 x 4783) + 1 = 9567.

|||||||||

{ 121 }

Which number fits in the next spot in this sequence?:

204, 817, 3269, 13077, 52309, ...?

||||

By observation and trial and error, we find that each term in this

sequence is obtained by multiplying the previous term by 4 and adding 1.

Therefore, the sixth term = (4 x 52309) + 1 = 209237.

||||||||

{ 122 }

Which number fits in the next spot in this sequence?:

9.6, 28.8, 86.4, 259.2, 777.6, ...?

||||

By taking the ratios of adjacent terms, we can see that this is a

Geometric Progression, with common ratio = 3. Therefore, the sixth term

can be calculated by multiplying the fifth term with the common ratio. It is

equal to 777.6 x 3 = 2332.8.

|||||||

{ 123 }

Which number fits in the next spot in this sequence?:

252, 505, 1011, 2023, 4047, ...?

||||

By observation and trial and error, we find that each term in this

sequence is obtained by multiplying the previous term by 2 and adding 1.

Therefore, the sixth term = (2 x 4047) + 1 = 8095.

|||||||

{ 124 }

Solve this: what comes next in this sequence?:

148, 188, 228, 268, 308, ...?

||||

By taking the differences between adjacent terms, we can see that this is

an Arithmetic Progression, with common difference = 40. Therefore, the

sixth term can be calculated by adding the common difference to the fifth

term. It is equal to 308 + 40 = 348.

|||||||||

{ 125 }

Which number fits in the next spot in this sequence?:

1.6, 6.4, 25.6, 102.4, 409.6, ...?

||||

By taking the ratios of adjacent terms, we can see that this is a

Geometric Progression, with common ratio = 4. Therefore, the sixth term

can be calculated by multiplying the fifth term with the common ratio. It is

equal to 409.6 x 4 = 1638.4.

|||||||||

{ 126 }

Solve this: what comes next in this sequence?:

221, 251, 281, 311, 341, ...?

||||

By taking the differences between adjacent terms, we can see that this is an Arithmetic Progression, with common difference = 30. Therefore, the sixth term can be calculated by adding the common difference to the fifth term. It is equal to 341 + 30 = 371.

|||||||||

{ 127 }

Which number fits in the next spot in this sequence?:

3.9, 15.6, 62.4, 249.6, 998.4, ...?

||||

By taking the ratios of adjacent terms, we can see that this is a Geometric Progression, with common ratio = 4. Therefore, the sixth term can be calculated by multiplying the fifth term with the common ratio. It is equal to 998.4 x 4 = 3993.6.

||||||||

{ 128 }

Solve this: what comes next in this sequence?:

197, 208, 219, 230, 241, ...?

||||

By taking the differences between adjacent terms, we can see that this is an Arithmetic Progression, with common difference = 11. Therefore, the

sixth term can be calculated by adding the common difference to the fifth term. It is equal to 241 + 11 = 252.

IIIIIIII

{ 129 }

Which number fits in the next spot in this sequence?:

210, 841, 3365, 13461, 53845, ...?

IIII

By observation and trial and error, we find that each term in this sequence is obtained by multiplying the previous term by 4 and adding 1. Therefore, the sixth term = (4 x 53845) + 1 = 215381.

IIIIIIII

{ 130 }

Which number fits in the next spot in this sequence?:

171, 343, 687, 1375, 2751, ...?

IIII

By observation and trial and error, we find that each term in this sequence is obtained by multiplying the previous term by 2 and adding 1. Therefore, the sixth term = (2 x 2751) + 1 = 5503.

IIIIIIII

{ 131 }

Solve this: what comes next in this sequence?:

95, 130, 165, 200, 235, ...?

||||

By taking the differences between adjacent terms, we can see that this is an Arithmetic Progression, with common difference = 35. Therefore, the sixth term can be calculated by adding the common difference to the fifth term. It is equal to 235 + 35 = 270.

|||||||||

{ 132 }

Solve this: what comes next in this sequence?:

270, 306, 342, 378, 414, ...?

||||

By taking the differences between adjacent terms, we can see that this is an Arithmetic Progression, with common difference = 36. Therefore, the sixth term can be calculated by adding the common difference to the fifth term. It is equal to 414 + 36 = 450.

|||||||||

{ 133 }

Which number fits in the next spot in this sequence?:

191, 383, 767, 1535, 3071, ...?

||||

By observation and trial and error, we find that each term in this sequence is obtained by multiplying the previous term by 2 and adding 1. Therefore, the sixth term = (2 x 3071) + 1 = 6143.

|||||||||

{ 134 }

Solve this: what comes next in this sequence?:

116, 154, 192, 230, 268, ...?

||||

By taking the differences between adjacent terms, we can see that this is an Arithmetic Progression, with common difference = 38. Therefore, the sixth term can be calculated by adding the common difference to the fifth term. It is equal to 268 + 38 = 306.

|||||||||

{ 135 }

Solve this: what comes next in this sequence?:

139, 173, 207, 241, 275, ...?

||||

By taking the differences between adjacent terms, we can see that this is an Arithmetic Progression, with common difference = 34. Therefore, the sixth term can be calculated by adding the common difference to the fifth

term. It is equal to 275 + 34 = 309.

llllllll

{ 136 }

Which number fits in the next spot in this sequence?:

299, 898, 2695, 8086, 24259, ...?

llll

By observation and trial and error, we find that each term in this

sequence is obtained by multiplying the previous term by 3 and adding 1.

Therefore, the sixth term = (3 x 24259) + 1 = 72778.

llllllll

{ 137 }

Which number fits in the next spot in this sequence?:

222, 667, 2002, 6007, 18022, ...?

llll

By observation and trial and error, we find that each term in this

sequence is obtained by multiplying the previous term by 3 and adding 1.

Therefore, the sixth term = (3 x 18022) + 1 = 54067.

llllllll

{ 138 }

Solve this: what comes next in this sequence?:

177, 222, 267, 312, 357, ...?

||||

By taking the differences between adjacent terms, we can see that this is

an Arithmetic Progression, with common difference = 45. Therefore, the

sixth term can be calculated by adding the common difference to the fifth

term. It is equal to 357 + 45 = 402.

|||||||||

{ 139 }

Which number fits in the next spot in this sequence?:

2.5, 5, 10, 20, 40, ...?

||||

By taking the ratios of adjacent terms, we can see that this is a

Geometric Progression, with common ratio = 2. Therefore, the sixth term

can be calculated by multiplying the fifth term with the common ratio. It is

equal to 40 x 2 = 80.

|||||||||

{ 140 }

Which number fits in the next spot in this sequence?:

1.3, 5.2, 20.8, 83.2, 332.8, ...?

IIII

By taking the ratios of adjacent terms, we can see that this is a

Geometric Progression, with common ratio = 4. Therefore, the sixth term

can be calculated by multiplying the fifth term with the common ratio. It is

equal to 332.8 x 4 = 1331.2.

IIIIIIII

{ 141 }

Which number fits in the next spot in this sequence?:

296, 593, 1187, 2375, 4751, ...?

IIII

By observation and trial and error, we find that each term in this

sequence is obtained by multiplying the previous term by 2 and adding 1.

Therefore, the sixth term = (2 x 4751) + 1 = 9503.

IIIIIIII

{ 142 }

Which number fits in the next spot in this sequence?:

166, 499, 1498, 4495, 13486, ...?

IIII

By observation and trial and error, we find that each term in this

sequence is obtained by multiplying the previous term by 3 and adding 1.

Therefore, the sixth term = (3 x 13486) + 1 = 40459.

|||||||||

{ 143 }

Solve this: what comes next in this sequence?:

119, 156, 193, 230, 267, ...?

||||

By taking the differences between adjacent terms, we can see that this is an Arithmetic Progression, with common difference = 37. Therefore, the sixth term can be calculated by adding the common difference to the fifth term. It is equal to 267 + 37 = 304.

|||||||||

{ 144 }

Solve this: what comes next in this sequence?:

202, 249, 296, 343, 390, ...?

||||

By taking the differences between adjacent terms, we can see that this is an Arithmetic Progression, with common difference = 47. Therefore, the sixth term can be calculated by adding the common difference to the fifth term. It is equal to 390 + 47 = 437.

|||||||||

{ 145 }

Which number fits in the next spot in this sequence?:

8.4, 25.2, 75.6, 226.8, 680.4, ...?

||||

By taking the ratios of adjacent terms, we can see that this is a

Geometric Progression, with common ratio = 3. Therefore, the sixth term

can be calculated by multiplying the fifth term with the common ratio. It is

equal to 680.4 x 3 = 2041.2.

|||||||||

{ 146 }

Which number fits in the next spot in this sequence?:

9.3, 27.9, 83.7, 251.1, 753.3, ...?

||||

By taking the ratios of adjacent terms, we can see that this is a

Geometric Progression, with common ratio = 3. Therefore, the sixth term

can be calculated by multiplying the fifth term with the common ratio. It is

equal to 753.3 x 3 = 2259.9.

|||||||||

{ 147 }

Solve this: what comes next in this sequence?:

170, 191, 212, 233, 254, ...?

||||

By taking the differences between adjacent terms, we can see that this is an Arithmetic Progression, with common difference = 21. Therefore, the sixth term can be calculated by adding the common difference to the fifth term. It is equal to 254 + 21 = 275.

|||||||||

{ 148 }

Which number fits in the next spot in this sequence?:

4.5, 18, 72, 288, 1152, ...?

||||

By taking the ratios of adjacent terms, we can see that this is a Geometric Progression, with common ratio = 4. Therefore, the sixth term can be calculated by multiplying the fifth term with the common ratio. It is equal to 1152 x 4 = 4608.

|||||||||

{ 149 }

Which number fits in the next spot in this sequence?:

3.7, 7.4, 14.8, 29.6, 59.2, ...?

||||

By taking the ratios of adjacent terms, we can see that this is a Geometric Progression, with common ratio = 2. Therefore, the sixth term

can be calculated by multiplying the fifth term with the common ratio. It is equal to 59.2 x 2 = 118.4.

|||||||||

{ 150 }

Solve this: what comes next in this sequence?:

208, 222, 236, 250, 264, ...?

|||||

By taking the differences between adjacent terms, we can see that this is an Arithmetic Progression, with common difference = 14. Therefore, the sixth term can be calculated by adding the common difference to the fifth term. It is equal to 264 + 14 = 278.

|||||||||

{ 151 }

Which number fits in the next spot in this sequence?:

7.3, 29.2, 116.8, 467.2, 1868.8, ...?

|||||

By taking the ratios of adjacent terms, we can see that this is a Geometric Progression, with common ratio = 4. Therefore, the sixth term can be calculated by multiplying the fifth term with the common ratio. It is equal to 1868.8 x 4 = 7475.2.

|||||||||

{ 152 }

Which number fits in the next spot in this sequence?:

6.6, 19.8, 59.4, 178.2, 534.6, ...?

||||

By taking the ratios of adjacent terms, we can see that this is a

Geometric Progression, with common ratio = 3. Therefore, the sixth term

can be calculated by multiplying the fifth term with the common ratio. It is

equal to 534.6 x 3 = 1603.8.

|||||||||

{ 153 }

Which number fits in the next spot in this sequence?:

5.6, 11.2, 22.4, 44.8, 89.6, ...?

||||

By taking the ratios of adjacent terms, we can see that this is a

Geometric Progression, with common ratio = 2. Therefore, the sixth term

can be calculated by multiplying the fifth term with the common ratio. It is

equal to 89.6 x 2 = 179.2.

|||||||||

{ 154 }

Solve this: what comes next in this sequence?:

207, 243, 279, 315, 351, ...?

||||

By taking the differences between adjacent terms, we can see that this is

an Arithmetic Progression, with common difference = 36. Therefore, the

sixth term can be calculated by adding the common difference to the fifth

term. It is equal to 351 + 36 = 387.

|||||||||

{ 155 }

Which number fits in the next spot in this sequence?:

5.2, 15.6, 46.8, 140.4, 421.2, ...?

||||

By taking the ratios of adjacent terms, we can see that this is a

Geometric Progression, with common ratio = 3. Therefore, the sixth term

can be calculated by multiplying the fifth term with the common ratio. It is

equal to 421.2 x 3 = 1263.6.

|||||||||

{ 156 }

Which number fits in the next spot in this sequence?:

256, 513, 1027, 2055, 4111, ...?

||||

By observation and trial and error, we find that each term in this

sequence is obtained by multiplying the previous term by 2 and adding 1.

Therefore, the sixth term = (2 x 4111) + 1 = 8223.

|||||||||

{ 157 }

Which number fits in the next spot in this sequence?:

6.3, 12.6, 25.2, 50.4, 100.8, ...?

||||

By taking the ratios of adjacent terms, we can see that this is a

Geometric Progression, with common ratio = 2. Therefore, the sixth term

can be calculated by multiplying the fifth term with the common ratio. It is

equal to 100.8 x 2 = 201.6.

|||||||||

{ 158 }

Solve this: what comes next in this sequence?:

205, 237, 269, 301, 333, ...?

||||

By taking the differences between adjacent terms, we can see that this is

an Arithmetic Progression, with common difference = 32. Therefore, the

sixth term can be calculated by adding the common difference to the fifth

term. It is equal to 333 + 32 = 365.

|||||||||

{ 159 }

Solve this: what comes next in this sequence?:

161, 191, 221, 251, 281, ...?

||||

By taking the differences between adjacent terms, we can see that this is

an Arithmetic Progression, with common difference = 30. Therefore, the

sixth term can be calculated by adding the common difference to the fifth

term. It is equal to 281 + 30 = 311.

|||||||||

{ 160 }

Which number fits in the next spot in this sequence?:

162, 487, 1462, 4387, 13162, ...?

||||

By observation and trial and error, we find that each term in this

sequence is obtained by multiplying the previous term by 3 and adding 1.

Therefore, the sixth term = (3 x 13162) + 1 = 39487.

|||||||||

{ 161 }

Which number fits in the next spot in this sequence?:

229, 459, 919, 1839, 3679, ...?

||||

By observation and trial and error, we find that each term in this sequence is obtained by multiplying the previous term by 2 and adding 1. Therefore, the sixth term = (2 x 3679) + 1 = 7359.

||||||||

{ 162 }

Which number fits in the next spot in this sequence?:

274, 823, 2470, 7411, 22234, ...?

||||

By observation and trial and error, we find that each term in this sequence is obtained by multiplying the previous term by 3 and adding 1. Therefore, the sixth term = (3 x 22234) + 1 = 66703.

||||||||

{ 163 }

Which number fits in the next spot in this sequence?:

1.2, 2.4, 4.8, 9.6, 19.2, ...?

||||

By taking the ratios of adjacent terms, we can see that this is a Geometric Progression, with common ratio = 2. Therefore, the sixth term can be calculated by multiplying the fifth term with the common ratio. It is equal to 19.2 x 2 = 38.4.

||||||||

{ 164 }

Which number fits in the next spot in this sequence?:

148, 593, 2373, 9493, 37973, ...?

||||

By observation and trial and error, we find that each term in this

sequence is obtained by multiplying the previous term by 4 and adding 1.

Therefore, the sixth term = (4 x 37973) + 1 = 151893.

|||||||||

{ 165 }

Solve this: what comes next in this sequence?:

121, 131, 141, 151, 161, ...?

||||

By taking the differences between adjacent terms, we can see that this is

an Arithmetic Progression, with common difference = 10. Therefore, the

sixth term can be calculated by adding the common difference to the fifth

term. It is equal to 161 + 10 = 171.

|||||||||

{ 166 }

Which number fits in the next spot in this sequence?:

155, 466, 1399, 4198, 12595, ...?

||||

By observation and trial and error, we find that each term in this sequence is obtained by multiplying the previous term by 3 and adding 1. Therefore, the sixth term = (3 x 12595) + 1 = 37786.

||||||||

{ 167 }

Which number fits in the next spot in this sequence?:

6.3, 18.9, 56.7, 170.1, 510.3, ...?

||||

By taking the ratios of adjacent terms, we can see that this is a Geometric Progression, with common ratio = 3. Therefore, the sixth term can be calculated by multiplying the fifth term with the common ratio. It is equal to 510.3 x 3 = 1530.9.

|||||||||

{ 168 }

Which number fits in the next spot in this sequence?:

2.1, 6.3, 18.9, 56.7, 170.1, ...?

||||

By taking the ratios of adjacent terms, we can see that this is a Geometric Progression, with common ratio = 3. Therefore, the sixth term can be calculated by multiplying the fifth term with the common ratio. It is equal to 170.1 x 3 = 510.3.

IIIIIIII

{ 169 }

Which number fits in the next spot in this sequence?:

277, 832, 2497, 7492, 22477, ...?

IIII

By observation and trial and error, we find that each term in this

sequence is obtained by multiplying the previous term by 3 and adding 1.

Therefore, the sixth term = (3 x 22477) + 1 = 67432.

IIIIIIII

{ 170 }

Which number fits in the next spot in this sequence?:

104, 417, 1669, 6677, 26709, ...?

IIII

By observation and trial and error, we find that each term in this

sequence is obtained by multiplying the previous term by 4 and adding 1.

Therefore, the sixth term = (4 x 26709) + 1 = 106837.

IIIIIIII

{ 171 }

Which number fits in the next spot in this sequence?:

9.9, 29.7, 89.1, 267.3, 801.9, ...?

||||

By taking the ratios of adjacent terms, we can see that this is a Geometric Progression, with common ratio = 3. Therefore, the sixth term can be calculated by multiplying the fifth term with the common ratio. It is equal to 801.9 x 3 = 2405.7.

|||||||||

{ 172 }

Which number fits in the next spot in this sequence?:

3.9, 7.8, 15.6, 31.2, 62.4, ...?

||||

By taking the ratios of adjacent terms, we can see that this is a Geometric Progression, with common ratio = 2. Therefore, the sixth term can be calculated by multiplying the fifth term with the common ratio. It is equal to 62.4 x 2 = 124.8.

|||||||||

{ 173 }

Which number fits in the next spot in this sequence?:

209, 837, 3349, 13397, 53589, ...?

||||

By observation and trial and error, we find that each term in this
sequence is obtained by multiplying the previous term by 4 and adding 1.
Therefore, the sixth term = (4 x 53589) + 1 = 214357.

||||||||

{ 174 }

Solve this: what comes next in this sequence?:

241, 260, 279, 298, 317, ...?

||||

By taking the differences between adjacent terms, we can see that this is
an Arithmetic Progression, with common difference = 19. Therefore, the
sixth term can be calculated by adding the common difference to the fifth
term. It is equal to 317 + 19 = 336.

|||||||||

{ 175 }

Which number fits in the next spot in this sequence?:

3.2, 12.8, 51.2, 204.8, 819.2, ...?

||||

By taking the ratios of adjacent terms, we can see that this is a
Geometric Progression, with common ratio = 4. Therefore, the sixth term
can be calculated by multiplying the fifth term with the common ratio. It is

equal to 819.2 x 4 = 3276.8.

||||||||

{ 176 }

Solve this: what comes next in this sequence?:

237, 274, 311, 348, 385, ...?

||||

By taking the differences between adjacent terms, we can see that this is

an Arithmetic Progression, with common difference = 37. Therefore, the

sixth term can be calculated by adding the common difference to the fifth

term. It is equal to 385 + 37 = 422.

||||||||

{ 177 }

Which number fits in the next spot in this sequence?:

1.5, 6, 24, 96, 384, ...?

||||

By taking the ratios of adjacent terms, we can see that this is a

Geometric Progression, with common ratio = 4. Therefore, the sixth term

can be calculated by multiplying the fifth term with the common ratio. It is

equal to 384 x 4 = 1536.

||||||||

{ 178 }

Which number fits in the next spot in this sequence?:

194, 583, 1750, 5251, 15754, ...?

||||

By observation and trial and error, we find that each term in this

sequence is obtained by multiplying the previous term by 3 and adding 1.

Therefore, the sixth term = (3 x 15754) + 1 = 47263.

|||||||||

{ 179 }

Solve this: what comes next in this sequence?:

158, 185, 212, 239, 266, ...?

||||

By taking the differences between adjacent terms, we can see that this is

an Arithmetic Progression, with common difference = 27. Therefore, the

sixth term can be calculated by adding the common difference to the fifth

term. It is equal to 266 + 27 = 293.

|||||||||

{ 180 }

Which number fits in the next spot in this sequence?:

3, 12, 48, 192, 768, ...?

||||

By taking the ratios of adjacent terms, we can see that this is a Geometric Progression, with common ratio = 4. Therefore, the sixth term can be calculated by multiplying the fifth term with the common ratio. It is equal to 768 x 4 = 3072.

|||||||||

{ 181 }

Which number fits in the next spot in this sequence?:

295, 886, 2659, 7978, 23935, ...?

||||

By observation and trial and error, we find that each term in this sequence is obtained by multiplying the previous term by 3 and adding 1. Therefore, the sixth term = (3 x 23935) + 1 = 71806.

|||||||||

{ 182 }

Which number fits in the next spot in this sequence?:

237, 712, 2137, 6412, 19237, ...?

||||

By observation and trial and error, we find that each term in this sequence is obtained by multiplying the previous term by 3 and adding 1. Therefore, the sixth term = (3 x 19237) + 1 = 57712.

|||||||||

{ 183 }

Which number fits in the next spot in this sequence?:

6.3, 12.6, 25.2, 50.4, 100.8, ...?

||||

By taking the ratios of adjacent terms, we can see that this is a

Geometric Progression, with common ratio = 2. Therefore, the sixth term

can be calculated by multiplying the fifth term with the common ratio. It is

equal to 100.8 x 2 = 201.6.

||||||||

{ 184 }

Solve this: what comes next in this sequence?:

130, 180, 230, 280, 330, ...?

||||

By taking the differences between adjacent terms, we can see that this is

an Arithmetic Progression, with common difference = 50. Therefore, the

sixth term can be calculated by adding the common difference to the fifth

term. It is equal to 330 + 50 = 380.

||||||||

{ 185 }

Solve this: what comes next in this sequence?:

163, 205, 247, 289, 331, ...?

||||

By taking the differences between adjacent terms, we can see that this is an Arithmetic Progression, with common difference = 42. Therefore, the sixth term can be calculated by adding the common difference to the fifth term. It is equal to 331 + 42 = 373.

|||||||||

{ 186 }

Solve this: what comes next in this sequence?:

173, 215, 257, 299, 341, ...?

||||

By taking the differences between adjacent terms, we can see that this is an Arithmetic Progression, with common difference = 42. Therefore, the sixth term can be calculated by adding the common difference to the fifth term. It is equal to 341 + 42 = 383.

|||||||||

{ 187 }

Which number fits in the next spot in this sequence?:

90, 181, 363, 727, 1455, ...?

||||

By observation and trial and error, we find that each term in this sequence is obtained by multiplying the previous term by 2 and adding 1.

Therefore, the sixth term = (2 x 1455) + 1 = 2911.

|||||||||

{ 188 }

Which number fits in the next spot in this sequence?:

6.3, 12.6, 25.2, 50.4, 100.8, ...?

||||

By taking the ratios of adjacent terms, we can see that this is a

Geometric Progression, with common ratio = 2. Therefore, the sixth term

can be calculated by multiplying the fifth term with the common ratio. It is

equal to 100.8 x 2 = 201.6.

|||||||||

{ 189 }

Solve this: what comes next in this sequence?:

248, 295, 342, 389, 436, ...?

||||

By taking the differences between adjacent terms, we can see that this is

an Arithmetic Progression, with common difference = 47. Therefore, the

sixth term can be calculated by adding the common difference to the fifth

term. It is equal to 436 + 47 = 483.

|||||||||

{ 190 }

Which number fits in the next spot in this sequence?:

171, 343, 687, 1375, 2751, ...?

||||

By observation and trial and error, we find that each term in this

sequence is obtained by multiplying the previous term by 2 and adding 1.

Therefore, the sixth term = (2 x 2751) + 1 = 5503.

|||||||||

{ 191 }

Which number fits in the next spot in this sequence?:

235, 706, 2119, 6358, 19075, ...?

||||

By observation and trial and error, we find that each term in this

sequence is obtained by multiplying the previous term by 3 and adding 1.

Therefore, the sixth term = (3 x 19075) + 1 = 57226.

|||||||||

{ 192 }

Which number fits in the next spot in this sequence?:

3.4, 6.8, 13.6, 27.2, 54.4, ...?

||||

By taking the ratios of adjacent terms, we can see that this is a

Geometric Progression, with common ratio = 2. Therefore, the sixth term can be calculated by multiplying the fifth term with the common ratio. It is equal to 54.4 x 2 = 108.8.

||||||||

{ 193 }

Which number fits in the next spot in this sequence?:

118, 473, 1893, 7573, 30293, ...?

||||

By observation and trial and error, we find that each term in this sequence is obtained by multiplying the previous term by 4 and adding 1. Therefore, the sixth term = (4 x 30293) + 1 = 121173.

||||||||

{ 194 }

Which number fits in the next spot in this sequence?:

4.8, 14.4, 43.2, 129.6, 388.8, ...?

||||

By taking the ratios of adjacent terms, we can see that this is a Geometric Progression, with common ratio = 3. Therefore, the sixth term can be calculated by multiplying the fifth term with the common ratio. It is equal to 388.8 x 3 = 1166.4.

||||||||

{ 195 }

Which number fits in the next spot in this sequence?:

5, 10, 20, 40, 80, ...?

||||

By taking the ratios of adjacent terms, we can see that this is a

Geometric Progression, with common ratio = 2. Therefore, the sixth term

can be calculated by multiplying the fifth term with the common ratio. It is

equal to 80 x 2 = 160.

|||||||||

{ 196 }

Which number fits in the next spot in this sequence?:

105, 316, 949, 2848, 8545, ...?

||||

By observation and trial and error, we find that each term in this

sequence is obtained by multiplying the previous term by 3 and adding 1.

Therefore, the sixth term = (3 x 8545) + 1 = 25636.

|||||||||

{ 197 }

Solve this: what comes next in this sequence?:

286, 332, 378, 424, 470, ...?

||||

By taking the differences between adjacent terms, we can see that this is an Arithmetic Progression, with common difference = 46. Therefore, the sixth term can be calculated by adding the common difference to the fifth term. It is equal to 470 + 46 = 516.

|||||||||

{ 198 }

Which number fits in the next spot in this sequence?:

93, 280, 841, 2524, 7573, ...?

||||

By observation and trial and error, we find that each term in this sequence is obtained by multiplying the previous term by 3 and adding 1. Therefore, the sixth term = (3 x 7573) + 1 = 22720.

|||||||||

{ 199 }

Solve this: what comes next in this sequence?:

135, 145, 155, 165, 175, ...?

||||

By taking the differences between adjacent terms, we can see that this is an Arithmetic Progression, with common difference = 10. Therefore, the sixth term can be calculated by adding the common difference to the fifth term. It is equal to 175 + 10 = 185.

||||||||

{ 200 }

Which number fits in the next spot in this sequence?:

8.1, 32.4, 129.6, 518.4, 2073.6, ...?

||||

By taking the ratios of adjacent terms, we can see that this is a

Geometric Progression, with common ratio = 4. Therefore, the sixth term

can be calculated by multiplying the fifth term with the common ratio. It is

equal to 2073.6 x 4 = 8294.4.

|||||||||

{ 201 }

Solve this: what comes next in this sequence?:

106, 134, 162, 190, 218, ...?

||||

By taking the differences between adjacent terms, we can see that this is

an Arithmetic Progression, with common difference = 28. Therefore, the

sixth term can be calculated by adding the common difference to the fifth

term. It is equal to 218 + 28 = 246.

|||||||||

{ 202 }

Which number fits in the next spot in this sequence?:

9.1, 27.3, 81.9, 245.7, 737.1, ...?

||||

By taking the ratios of adjacent terms, we can see that this is a

Geometric Progression, with common ratio = 3. Therefore, the sixth term

can be calculated by multiplying the fifth term with the common ratio. It is

equal to 737.1 x 3 = 2211.3.

|||||||||

{ 203 }

Solve this: what comes next in this sequence?:

120, 135, 150, 165, 180, ...?

||||

By taking the differences between adjacent terms, we can see that this is

an Arithmetic Progression, with common difference = 15. Therefore, the

sixth term can be calculated by adding the common difference to the fifth

term. It is equal to 180 + 15 = 195.

|||||||||

{ 204 }

Which number fits in the next spot in this sequence?:

282, 565, 1131, 2263, 4527, ...?

||||

By observation and trial and error, we find that each term in this sequence is obtained by multiplying the previous term by 2 and adding 1. Therefore, the sixth term = (2 x 4527) + 1 = 9055.

|||||||||

{ 205 }

Which number fits in the next spot in this sequence?:

9.6, 19.2, 38.4, 76.8, 153.6, ...?

||||

By taking the ratios of adjacent terms, we can see that this is a Geometric Progression, with common ratio = 2. Therefore, the sixth term can be calculated by multiplying the fifth term with the common ratio. It is equal to 153.6 x 2 = 307.2.

||||||||

{ 206 }

Solve this: what comes next in this sequence?:

235, 278, 321, 364, 407, ...?

||||

By taking the differences between adjacent terms, we can see that this is an Arithmetic Progression, with common difference = 43. Therefore, the sixth term can be calculated by adding the common difference to the fifth

term. It is equal to 407 + 43 = 450.

||||||||

{ 207 }

Which number fits in the next spot in this sequence?:

5.8, 11.6, 23.2, 46.4, 92.8000000000001, ...?

||||

By taking the ratios of adjacent terms, we can see that this is a

Geometric Progression, with common ratio = 2. Therefore, the sixth term

can be calculated by multiplying the fifth term with the common ratio. It is

equal to 92.8000000000001 x 2 = 185.6.

||||||||

{ 208 }

Which number fits in the next spot in this sequence?:

237, 949, 3797, 15189, 60757, ...?

||||

By observation and trial and error, we find that each term in this

sequence is obtained by multiplying the previous term by 4 and adding 1.

Therefore, the sixth term = (4 x 60757) + 1 = 243029.

||||||||

{ 209 }

Which number fits in the next spot in this sequence?:

3.5, 7, 14, 28, 56, ...?

||||

By taking the ratios of adjacent terms, we can see that this is a

Geometric Progression, with common ratio = 2. Therefore, the sixth term

can be calculated by multiplying the fifth term with the common ratio. It is

equal to 56 x 2 = 112.

|||||||||

{ 210 }

Which number fits in the next spot in this sequence?:

117, 469, 1877, 7509, 30037, ...?

||||

By observation and trial and error, we find that each term in this

sequence is obtained by multiplying the previous term by 4 and adding 1.

Therefore, the sixth term = (4 x 30037) + 1 = 120149.

|||||||||

{ 211 }

Solve this: what comes next in this sequence?:

142, 180, 218, 256, 294, ...?

||||

By taking the differences between adjacent terms, we can see that this is an Arithmetic Progression, with common difference = 38. Therefore, the sixth term can be calculated by adding the common difference to the fifth term. It is equal to 294 + 38 = 332.

||||||||

{ 212 }

Solve this: what comes next in this sequence?:

281, 322, 363, 404, 445, ...?

||||

By taking the differences between adjacent terms, we can see that this is an Arithmetic Progression, with common difference = 41. Therefore, the sixth term can be calculated by adding the common difference to the fifth term. It is equal to 445 + 41 = 486.

|||||||||

{ 213 }

Solve this: what comes next in this sequence?:

228, 262, 296, 330, 364, ...?

||||

By taking the differences between adjacent terms, we can see that this is an Arithmetic Progression, with common difference = 34. Therefore, the sixth term can be calculated by adding the common difference to the fifth

term. It is equal to 364 + 34 = 398.

|||||||||

{ 214 }

Which number fits in the next spot in this sequence?:

2.2, 6.6, 19.8, 59.4, 178.2, ...?

||||

By taking the ratios of adjacent terms, we can see that this is a

Geometric Progression, with common ratio = 3. Therefore, the sixth term

can be calculated by multiplying the fifth term with the common ratio. It is

equal to 178.2 x 3 = 534.600000000001.

|||||||||

{ 215 }

Which number fits in the next spot in this sequence?:

286, 1145, 4581, 18325, 73301, ...?

||||

By observation and trial and error, we find that each term in this

sequence is obtained by multiplying the previous term by 4 and adding 1.

Therefore, the sixth term = (4 x 73301) + 1 = 293205.

|||||||||

{ 216 }

Solve this: what comes next in this sequence?:

232, 271, 310, 349, 388, ...?

||||

By taking the differences between adjacent terms, we can see that this is an Arithmetic Progression, with common difference = 39. Therefore, the sixth term can be calculated by adding the common difference to the fifth term. It is equal to 388 + 39 = 427.

|||||||||

{ 217 }

Which number fits in the next spot in this sequence?:

4.5, 9, 18, 36, 72, ...?

||||

By taking the ratios of adjacent terms, we can see that this is a Geometric Progression, with common ratio = 2. Therefore, the sixth term can be calculated by multiplying the fifth term with the common ratio. It is equal to 72 x 2 = 144.

|||||||||

{ 218 }

Which number fits in the next spot in this sequence?:

266, 533, 1067, 2135, 4271, ...?

||||

By observation and trial and error, we find that each term in this sequence is obtained by multiplying the previous term by 2 and adding 1. Therefore, the sixth term = (2 x 4271) + 1 = 8543.

|||||||||

{ 219 }

Solve this: what comes next in this sequence?:

99, 116, 133, 150, 167, ...?

||||

By taking the differences between adjacent terms, we can see that this is an Arithmetic Progression, with common difference = 17. Therefore, the sixth term can be calculated by adding the common difference to the fifth term. It is equal to 167 + 17 = 184.

||||||||

{ 220 }

Solve this: what comes next in this sequence?:

286, 303, 320, 337, 354, ...?

||||

By taking the differences between adjacent terms, we can see that this is an Arithmetic Progression, with common difference = 17. Therefore, the sixth term can be calculated by adding the common difference to the fifth

term. It is equal to 354 + 17 = 371.

||||||||

{ 221 }

Which number fits in the next spot in this sequence?:

7.3, 14.6, 29.2, 58.4, 116.8, ...?

||||

By taking the ratios of adjacent terms, we can see that this is a

Geometric Progression, with common ratio = 2. Therefore, the sixth term

can be calculated by multiplying the fifth term with the common ratio. It is

equal to 116.8 x 2 = 233.6.

||||||||

{ 222 }

Which number fits in the next spot in this sequence?:

113, 227, 455, 911, 1823, ...?

||||

By observation and trial and error, we find that each term in this

sequence is obtained by multiplying the previous term by 2 and adding 1.

Therefore, the sixth term = (2 x 1823) + 1 = 3647.

||||||||

{ 223 }

Which number fits in the next spot in this sequence?:

163, 653, 2613, 10453, 41813, ...?

||||

By observation and trial and error, we find that each term in this

sequence is obtained by multiplying the previous term by 4 and adding 1.

Therefore, the sixth term = (4 x 41813) + 1 = 167253.

|||||||||

{ 224 }

Which number fits in the next spot in this sequence?:

284, 853, 2560, 7681, 23044, ...?

||||

By observation and trial and error, we find that each term in this

sequence is obtained by multiplying the previous term by 3 and adding 1.

Therefore, the sixth term = (3 x 23044) + 1 = 69133.

|||||||||

{ 225 }

Which number fits in the next spot in this sequence?:

110, 331, 994, 2983, 8950, ...?

||||

By observation and trial and error, we find that each term in this

sequence is obtained by multiplying the previous term by 3 and adding 1.

Therefore, the sixth term = (3 x 8950) + 1 = 26851.

||||||||

{ 226 }

Which number fits in the next spot in this sequence?:

236, 709, 2128, 6385, 19156, ...?

||||

By observation and trial and error, we find that each term in this

sequence is obtained by multiplying the previous term by 3 and adding 1.

Therefore, the sixth term = (3 x 19156) + 1 = 57469.

||||||||

{ 227 }

Which number fits in the next spot in this sequence?:

8.7, 34.8, 139.2, 556.8, 2227.2, ...?

||||

By taking the ratios of adjacent terms, we can see that this is a

Geometric Progression, with common ratio = 4. Therefore, the sixth term

can be calculated by multiplying the fifth term with the common ratio. It is

equal to 2227.2 x 4 = 8908.8.

||||||||

{ 228 }

Solve this: what comes next in this sequence?:

109, 159, 209, 259, 309, ...?

||||

By taking the differences between adjacent terms, we can see that this is an Arithmetic Progression, with common difference = 50. Therefore, the sixth term can be calculated by adding the common difference to the fifth term. It is equal to 309 + 50 = 359.

|||||||||

{ 229 }

Which number fits in the next spot in this sequence?:

218, 873, 3493, 13973, 55893, ...?

||||

By observation and trial and error, we find that each term in this sequence is obtained by multiplying the previous term by 4 and adding 1. Therefore, the sixth term = (4 x 55893) + 1 = 223573.

|||||||||

{ 230 }

Solve this: what comes next in this sequence?:

160, 206, 252, 298, 344, ...?

||||

By taking the differences between adjacent terms, we can see that this is an Arithmetic Progression, with common difference = 46. Therefore, the sixth term can be calculated by adding the common difference to the fifth term. It is equal to 344 + 46 = 390.

||||||||

{ 231 }

Which number fits in the next spot in this sequence?:

253, 507, 1015, 2031, 4063, ...?

||||

By observation and trial and error, we find that each term in this sequence is obtained by multiplying the previous term by 2 and adding 1. Therefore, the sixth term = (2 x 4063) + 1 = 8127.

||||||||

{ 232 }

Which number fits in the next spot in this sequence?:

167, 669, 2677, 10709, 42837, ...?

||||

By observation and trial and error, we find that each term in this sequence is obtained by multiplying the previous term by 4 and adding 1. Therefore, the sixth term = (4 x 42837) + 1 = 171349.

||||||||

{ 233 }

Which number fits in the next spot in this sequence?:

4.1, 8.2, 16.4, 32.8, 65.6, ...?

||||

By taking the ratios of adjacent terms, we can see that this is a

Geometric Progression, with common ratio = 2. Therefore, the sixth term

can be calculated by multiplying the fifth term with the common ratio. It is

equal to 65.6 x 2 = 131.2.

|||||||||

{ 234 }

Solve this: what comes next in this sequence?:

251, 299, 347, 395, 443, ...?

||||

By taking the differences between adjacent terms, we can see that this is

an Arithmetic Progression, with common difference = 48. Therefore, the

sixth term can be calculated by adding the common difference to the fifth

term. It is equal to 443 + 48 = 491.

|||||||||

{ 235 }

Which number fits in the next spot in this sequence?:

253, 760, 2281, 6844, 20533, ...?

||||

By observation and trial and error, we find that each term in this sequence is obtained by multiplying the previous term by 3 and adding 1. Therefore, the sixth term = (3 x 20533) + 1 = 61600.

||||||||

{ 236 }

Solve this: what comes next in this sequence?:

98, 123, 148, 173, 198, ...?

||||

By taking the differences between adjacent terms, we can see that this is an Arithmetic Progression, with common difference = 25. Therefore, the sixth term can be calculated by adding the common difference to the fifth term. It is equal to 198 + 25 = 223.

|||||||||

{ 237 }

Which number fits in the next spot in this sequence?:

4.2, 8.4, 16.8, 33.6, 67.2, ...?

||||

By taking the ratios of adjacent terms, we can see that this is a Geometric Progression, with common ratio = 2. Therefore, the sixth term can be calculated by multiplying the fifth term with the common ratio. It is

equal to 67.2 x 2 = 134.4.

||||||||

{ 238 }

Solve this: what comes next in this sequence?:

224, 271, 318, 365, 412, ...?

||||

By taking the differences between adjacent terms, we can see that this is

an Arithmetic Progression, with common difference = 47. Therefore, the

sixth term can be calculated by adding the common difference to the fifth

term. It is equal to 412 + 47 = 459.

||||||||

{ 239 }

Solve this: what comes next in this sequence?:

141, 174, 207, 240, 273, ...?

||||

By taking the differences between adjacent terms, we can see that this is

an Arithmetic Progression, with common difference = 33. Therefore, the

sixth term can be calculated by adding the common difference to the fifth

term. It is equal to 273 + 33 = 306.

||||||||

{ 240 }

Which number fits in the next spot in this sequence?:

8.1, 24.3, 72.9, 218.7, 656.1, ...?

||||

By taking the ratios of adjacent terms, we can see that this is a

Geometric Progression, with common ratio = 3. Therefore, the sixth term

can be calculated by multiplying the fifth term with the common ratio. It is

equal to 656.1 x 3 = 1968.3.

|||||||||

{ 241 }

Which number fits in the next spot in this sequence?:

90, 181, 363, 727, 1455, ...?

||||

By observation and trial and error, we find that each term in this

sequence is obtained by multiplying the previous term by 2 and adding 1.

Therefore, the sixth term = (2 x 1455) + 1 = 2911.

|||||||||

{ 242 }

Solve this: what comes next in this sequence?:

285, 323, 361, 399, 437, ...?

||||

By taking the differences between adjacent terms, we can see that this is an Arithmetic Progression, with common difference = 38. Therefore, the sixth term can be calculated by adding the common difference to the fifth term. It is equal to 437 + 38 = 475.

||||||||

{ 243 }

Which number fits in the next spot in this sequence?:

2.1, 4.2, 8.4, 16.8, 33.6, ...?

||||

By taking the ratios of adjacent terms, we can see that this is a Geometric Progression, with common ratio = 2. Therefore, the sixth term can be calculated by multiplying the fifth term with the common ratio. It is equal to 33.6 x 2 = 67.2.

||||||||

{ 244 }

Solve this: what comes next in this sequence?:

129, 176, 223, 270, 317, ...?

||||

By taking the differences between adjacent terms, we can see that this is an Arithmetic Progression, with common difference = 47. Therefore, the sixth term can be calculated by adding the common difference to the fifth

term. It is equal to 317 + 47 = 364.

||||||||

{ 245 }

Which number fits in the next spot in this sequence?:

91, 365, 1461, 5845, 23381, ...?

||||

By observation and trial and error, we find that each term in this

sequence is obtained by multiplying the previous term by 4 and adding 1.

Therefore, the sixth term = (4 x 23381) + 1 = 93525.

||||||||

{ 246 }

Which number fits in the next spot in this sequence?:

97, 195, 391, 783, 1567, ...?

||||

By observation and trial and error, we find that each term in this

sequence is obtained by multiplying the previous term by 2 and adding 1.

Therefore, the sixth term = (2 x 1567) + 1 = 3135.

||||||||

{ 247 }

Solve this: what comes next in this sequence?:

148, 160, 172, 184, 196, ...?

||||

By taking the differences between adjacent terms, we can see that this is

an Arithmetic Progression, with common difference = 12. Therefore, the

sixth term can be calculated by adding the common difference to the fifth

term. It is equal to 196 + 12 = 208.

|||||||||

{ 248 }

Which number fits in the next spot in this sequence?:

8.6, 17.2, 34.4, 68.8, 137.6, ...?

||||

By taking the ratios of adjacent terms, we can see that this is a

Geometric Progression, with common ratio = 2. Therefore, the sixth term

can be calculated by multiplying the fifth term with the common ratio. It is

equal to 137.6 x 2 = 275.2.

|||||||||

{ 249 }

Which number fits in the next spot in this sequence?:

144, 433, 1300, 3901, 11704, ...?

||||

By observation and trial and error, we find that each term in this sequence is obtained by multiplying the previous term by 3 and adding 1. Therefore, the sixth term = (3 x 11704) + 1 = 35113.

||||||||

{ 250 }

Which number fits in the next spot in this sequence?:

261, 1045, 4181, 16725, 66901, ...?

||||

By observation and trial and error, we find that each term in this sequence is obtained by multiplying the previous term by 4 and adding 1. Therefore, the sixth term = (4 x 66901) + 1 = 267605.

|||||||||

{ 251 }

Solve this: what comes next in this sequence?:

216, 250, 284, 318, 352, ...?

||||

By taking the differences between adjacent terms, we can see that this is an Arithmetic Progression, with common difference = 34. Therefore, the sixth term can be calculated by adding the common difference to the fifth term. It is equal to 352 + 34 = 386.

IIIIIIII

{ 252 }

Which number fits in the next spot in this sequence?:

7.9, 23.7, 71.1, 213.3, 639.900000000001, ...?

IIII

By taking the ratios of adjacent terms, we can see that this is a

Geometric Progression, with common ratio = 3. Therefore, the sixth term

can be calculated by multiplying the fifth term with the common ratio. It is

equal to 639.900000000001 x 3 = 1919.7.

IIIIIIII

{ 253 }

Which number fits in the next spot in this sequence?:

9.9, 19.8, 39.6, 79.2, 158.4, ...?

IIII

By taking the ratios of adjacent terms, we can see that this is a

Geometric Progression, with common ratio = 2. Therefore, the sixth term

can be calculated by multiplying the fifth term with the common ratio. It is

equal to 158.4 x 2 = 316.8.

IIIIIIII

{ 254 }

Which number fits in the next spot in this sequence?:

293, 880, 2641, 7924, 23773, ...?

||||

By observation and trial and error, we find that each term in this
sequence is obtained by multiplying the previous term by 3 and adding 1.
Therefore, the sixth term = (3 x 23773) + 1 = 71320.

|||||||||

{ 255 }

Solve this: what comes next in this sequence?:

286, 329, 372, 415, 458, ...?

||||

By taking the differences between adjacent terms, we can see that this is
an Arithmetic Progression, with common difference = 43. Therefore, the
sixth term can be calculated by adding the common difference to the fifth
term. It is equal to 458 + 43 = 501.

|||||||||

{ 256 }

Solve this: what comes next in this sequence?:

290, 323, 356, 389, 422, ...?

||||

By taking the differences between adjacent terms, we can see that this is an Arithmetic Progression, with common difference = 33. Therefore, the sixth term can be calculated by adding the common difference to the fifth term. It is equal to 422 + 33 = 455.

|||||||||

{ 257 }

Solve this: what comes next in this sequence?:

217, 242, 267, 292, 317, ...?

||||

By taking the differences between adjacent terms, we can see that this is an Arithmetic Progression, with common difference = 25. Therefore, the sixth term can be calculated by adding the common difference to the fifth term. It is equal to 317 + 25 = 342.

|||||||||

{ 258 }

Which number fits in the next spot in this sequence?:

8.7, 26.1, 78.3, 234.9, 704.7, ...?

||||

By taking the ratios of adjacent terms, we can see that this is a Geometric Progression, with common ratio = 3. Therefore, the sixth term can be calculated by multiplying the fifth term with the common ratio. It is

equal to 704.7 x 3 = 2114.1.

|||||||||

{ 259 }

Which number fits in the next spot in this sequence?:

234, 937, 3749, 14997, 59989, ...?

||||

By observation and trial and error, we find that each term in this

sequence is obtained by multiplying the previous term by 4 and adding 1.

Therefore, the sixth term = (4 x 59989) + 1 = 239957.

|||||||||

{ 260 }

Which number fits in the next spot in this sequence?:

294, 1177, 4709, 18837, 75349, ...?

||||

By observation and trial and error, we find that each term in this

sequence is obtained by multiplying the previous term by 4 and adding 1.

Therefore, the sixth term = (4 x 75349) + 1 = 301397.

|||||||||

{ 261 }

Which number fits in the next spot in this sequence?:

8.9, 35.6, 142.4, 569.6, 2278.4, ...?

||||

By taking the ratios of adjacent terms, we can see that this is a

Geometric Progression, with common ratio = 4. Therefore, the sixth term

can be calculated by multiplying the fifth term with the common ratio. It is

equal to 2278.4 x 4 = 9113.6.

|||||||||

{ 262 }

Which number fits in the next spot in this sequence?:

214, 857, 3429, 13717, 54869, ...?

||||

By observation and trial and error, we find that each term in this

sequence is obtained by multiplying the previous term by 4 and adding 1.

Therefore, the sixth term = (4 x 54869) + 1 = 219477.

|||||||||

{ 263 }

Solve this: what comes next in this sequence?:

294, 336, 378, 420, 462, ...?

||||

By taking the differences between adjacent terms, we can see that this is an Arithmetic Progression, with common difference = 42. Therefore, the sixth term can be calculated by adding the common difference to the fifth term. It is equal to 462 + 42 = 504.

|||||||

{ 264 }

Solve this: what comes next in this sequence?:

210, 224, 238, 252, 266, ...?

||||

By taking the differences between adjacent terms, we can see that this is an Arithmetic Progression, with common difference = 14. Therefore, the sixth term can be calculated by adding the common difference to the fifth term. It is equal to 266 + 14 = 280.

|||||||

{ 265 }

Which number fits in the next spot in this sequence?:

3.6, 10.8, 32.4, 97.2, 291.6, ...?

||||

By taking the ratios of adjacent terms, we can see that this is a Geometric Progression, with common ratio = 3. Therefore, the sixth term can be calculated by multiplying the fifth term with the common ratio. It is

equal to 291.6 x 3 = 874.800000000001.

|||||||||

{ 266 }

Solve this: what comes next in this sequence?:

160, 209, 258, 307, 356, ...?

||||

By taking the differences between adjacent terms, we can see that this is

an Arithmetic Progression, with common difference = 49. Therefore, the

sixth term can be calculated by adding the common difference to the fifth

term. It is equal to 356 + 49 = 405.

|||||||||

{ 267 }

Which number fits in the next spot in this sequence?:

189, 379, 759, 1519, 3039, ...?

||||

By observation and trial and error, we find that each term in this

sequence is obtained by multiplying the previous term by 2 and adding 1.

Therefore, the sixth term = (2 x 3039) + 1 = 6079.

|||||||||

{ 268 }

Which number fits in the next spot in this sequence?:

221, 664, 1993, 5980, 17941, ...?

||||

By observation and trial and error, we find that each term in this
sequence is obtained by multiplying the previous term by 3 and adding 1.
Therefore, the sixth term = (3 x 17941) + 1 = 53824.

|||||||||

{ 269 }

Which number fits in the next spot in this sequence?:

7.6, 22.8, 68.4, 205.2, 615.6, ...?

||||

By taking the ratios of adjacent terms, we can see that this is a
Geometric Progression, with common ratio = 3. Therefore, the sixth term
can be calculated by multiplying the fifth term with the common ratio. It is
equal to 615.6 x 3 = 1846.8.

||||||||

{ 270 }

Which number fits in the next spot in this sequence?:

130, 261, 523, 1047, 2095, ...?

||||

By observation and trial and error, we find that each term in this

sequence is obtained by multiplying the previous term by 2 and adding 1.

Therefore, the sixth term = (2 x 2095) + 1 = 4191.

||||||||

{ 271 }

Which number fits in the next spot in this sequence?:

8.1, 16.2, 32.4, 64.8, 129.6, ...?

||||

By taking the ratios of adjacent terms, we can see that this is a

Geometric Progression, with common ratio = 2. Therefore, the sixth term

can be calculated by multiplying the fifth term with the common ratio. It is

equal to 129.6 x 2 = 259.2.

||||||||

{ 272 }

Which number fits in the next spot in this sequence?:

140, 281, 563, 1127, 2255, ...?

||||

By observation and trial and error, we find that each term in this

sequence is obtained by multiplying the previous term by 2 and adding 1.

Therefore, the sixth term = (2 x 2255) + 1 = 4511.

||||||||

{ 273 }

Which number fits in the next spot in this sequence?:

7.1, 14.2, 28.4, 56.8, 113.6, ...?

||||

By taking the ratios of adjacent terms, we can see that this is a

Geometric Progression, with common ratio = 2. Therefore, the sixth term

can be calculated by multiplying the fifth term with the common ratio. It is

equal to 113.6 x 2 = 227.2.

|||||||||

{ 274 }

Which number fits in the next spot in this sequence?:

7.4, 29.6, 118.4, 473.6, 1894.4, ...?

||||

By taking the ratios of adjacent terms, we can see that this is a

Geometric Progression, with common ratio = 4. Therefore, the sixth term

can be calculated by multiplying the fifth term with the common ratio. It is

equal to 1894.4 x 4 = 7577.6.

|||||||||

{ 275 }

Which number fits in the next spot in this sequence?:

2.3, 9.2, 36.8, 147.2, 588.8, ...?

||||

By taking the ratios of adjacent terms, we can see that this is a

Geometric Progression, with common ratio = 4. Therefore, the sixth term

can be calculated by multiplying the fifth term with the common ratio. It is

equal to 588.8 x 4 = 2355.2.

|||||||||

{ 276 }

Which number fits in the next spot in this sequence?:

125, 251, 503, 1007, 2015, ...?

||||

By observation and trial and error, we find that each term in this

sequence is obtained by multiplying the previous term by 2 and adding 1.

Therefore, the sixth term = (2 x 2015) + 1 = 4031.

|||||||||

{ 277 }

Solve this: what comes next in this sequence?:

229, 251, 273, 295, 317, ...?

||||

By taking the differences between adjacent terms, we can see that this is

an Arithmetic Progression, with common difference = 22. Therefore, the

sixth term can be calculated by adding the common difference to the fifth

term. It is equal to 317 + 22 = 339.

|||||||||

{ 278 }

Which number fits in the next spot in this sequence?:

288, 1153, 4613, 18453, 73813, ...?

||||

By observation and trial and error, we find that each term in this

sequence is obtained by multiplying the previous term by 4 and adding 1.

Therefore, the sixth term = (4 x 73813) + 1 = 295253.

|||||||||

{ 279 }

Which number fits in the next spot in this sequence?:

102, 307, 922, 2767, 8302, ...?

||||

By observation and trial and error, we find that each term in this

sequence is obtained by multiplying the previous term by 3 and adding 1.

Therefore, the sixth term = (3 x 8302) + 1 = 24907.

|||||||||

{ 280 }

Which number fits in the next spot in this sequence?:

4.8, 9.6, 19.2, 38.4, 76.8, ...?

||||

By taking the ratios of adjacent terms, we can see that this is a

Geometric Progression, with common ratio = 2. Therefore, the sixth term

can be calculated by multiplying the fifth term with the common ratio. It is

equal to 76.8 x 2 = 153.6.

|||||||||

{ 281 }

Which number fits in the next spot in this sequence?:

257, 515, 1031, 2063, 4127, ...?

||||

By observation and trial and error, we find that each term in this

sequence is obtained by multiplying the previous term by 2 and adding 1.

Therefore, the sixth term = (2 x 4127) + 1 = 8255.

|||||||||

{ 282 }

Which number fits in the next spot in this sequence?:

177, 355, 711, 1423, 2847, ...?

||||

By observation and trial and error, we find that each term in this sequence is obtained by multiplying the previous term by 2 and adding 1. Therefore, the sixth term = (2 x 2847) + 1 = 5695.

|||||||||

{ 283 }

Which number fits in the next spot in this sequence?:

8.7, 34.8, 139.2, 556.8, 2227.2, ...?

|||||

By taking the ratios of adjacent terms, we can see that this is a Geometric Progression, with common ratio = 4. Therefore, the sixth term can be calculated by multiplying the fifth term with the common ratio. It is equal to 2227.2 x 4 = 8908.8.

|||||||||

{ 284 }

Which number fits in the next spot in this sequence?:

8.7, 17.4, 34.8, 69.6, 139.2, ...?

|||||

By taking the ratios of adjacent terms, we can see that this is a Geometric Progression, with common ratio = 2. Therefore, the sixth term can be calculated by multiplying the fifth term with the common ratio. It is equal to 139.2 x 2 = 278.4.

|||||||||

{ 285 }

Which number fits in the next spot in this sequence?:

8.2, 32.8, 131.2, 524.8, 2099.2, ...?

||||

By taking the ratios of adjacent terms, we can see that this is a

Geometric Progression, with common ratio = 4. Therefore, the sixth term

can be calculated by multiplying the fifth term with the common ratio. It is

equal to 2099.2 x 4 = 8396.8.

|||||||||

{ 286 }

Which number fits in the next spot in this sequence?:

201, 604, 1813, 5440, 16321, ...?

||||

By observation and trial and error, we find that each term in this

sequence is obtained by multiplying the previous term by 3 and adding 1.

Therefore, the sixth term = (3 x 16321) + 1 = 48964.

|||||||||

{ 287 }

Which number fits in the next spot in this sequence?:

3.8, 11.4, 34.2, 102.6, 307.8, ...?

||||

By taking the ratios of adjacent terms, we can see that this is a

Geometric Progression, with common ratio = 3. Therefore, the sixth term

can be calculated by multiplying the fifth term with the common ratio. It is

equal to 307.8 x 3 = 923.4.

|||||||||

{ 288 }

Which number fits in the next spot in this sequence?:

115, 461, 1845, 7381, 29525, ...?

||||

By observation and trial and error, we find that each term in this

sequence is obtained by multiplying the previous term by 4 and adding 1.

Therefore, the sixth term = (4 x 29525) + 1 = 118101.

|||||||||

{ 289 }

Solve this: what comes next in this sequence?:

104, 120, 136, 152, 168, ...?

||||

By taking the differences between adjacent terms, we can see that this is an Arithmetic Progression, with common difference = 16. Therefore, the sixth term can be calculated by adding the common difference to the fifth term. It is equal to 168 + 16 = 184.

||||||||

{ 290 }

Solve this: what comes next in this sequence?:

108, 149, 190, 231, 272, ...?

||||

By taking the differences between adjacent terms, we can see that this is an Arithmetic Progression, with common difference = 41. Therefore, the sixth term can be calculated by adding the common difference to the fifth term. It is equal to 272 + 41 = 313.

||||||||

{ 291 }

Solve this: what comes next in this sequence?:

211, 224, 237, 250, 263, ...?

||||

By taking the differences between adjacent terms, we can see that this is an Arithmetic Progression, with common difference = 13. Therefore, the sixth term can be calculated by adding the common difference to the fifth

term. It is equal to 263 + 13 = 276.

|||||||||

{ 292 }

Solve this: what comes next in this sequence?:

174, 193, 212, 231, 250, ...?

||||

By taking the differences between adjacent terms, we can see that this is

an Arithmetic Progression, with common difference = 19. Therefore, the

sixth term can be calculated by adding the common difference to the fifth

term. It is equal to 250 + 19 = 269.

|||||||||

{ 293 }

Solve this: what comes next in this sequence?:

300, 315, 330, 345, 360, ...?

||||

By taking the differences between adjacent terms, we can see that this is

an Arithmetic Progression, with common difference = 15. Therefore, the

sixth term can be calculated by adding the common difference to the fifth

term. It is equal to 360 + 15 = 375.

|||||||||

{ 294 }

Which number fits in the next spot in this sequence?:

5.8, 23.2, 92.8, 371.2, 1484.8, ...?

||||

By taking the ratios of adjacent terms, we can see that this is a

Geometric Progression, with common ratio = 4. Therefore, the sixth term

can be calculated by multiplying the fifth term with the common ratio. It is

equal to 1484.8 x 4 = 5939.2.

|||||||||

{ 295 }

Which number fits in the next spot in this sequence?:

8.1, 24.3, 72.9, 218.7, 656.1, ...?

||||

By taking the ratios of adjacent terms, we can see that this is a

Geometric Progression, with common ratio = 3. Therefore, the sixth term

can be calculated by multiplying the fifth term with the common ratio. It is

equal to 656.1 x 3 = 1968.3.

|||||||||

{ 296 }

Which number fits in the next spot in this sequence?:

7.9, 15.8, 31.6, 63.2, 126.4, ...?

||||

By taking the ratios of adjacent terms, we can see that this is a

Geometric Progression, with common ratio = 2. Therefore, the sixth term

can be calculated by multiplying the fifth term with the common ratio. It is

equal to 126.4 x 2 = 252.8.

|||||||||

{ 297 }

Which number fits in the next spot in this sequence?:

9.6, 38.4, 153.6, 614.4, 2457.6, ...?

||||

By taking the ratios of adjacent terms, we can see that this is a

Geometric Progression, with common ratio = 4. Therefore, the sixth term

can be calculated by multiplying the fifth term with the common ratio. It is

equal to 2457.6 x 4 = 9830.4.

|||||||||

{ 298 }

Which number fits in the next spot in this sequence?:

8.3, 24.9, 74.7, 224.1, 672.3, ...?

||||

By taking the ratios of adjacent terms, we can see that this is a

Geometric Progression, with common ratio = 3. Therefore, the sixth term

can be calculated by multiplying the fifth term with the common ratio. It is equal to 672.3 x 3 = 2016.9.

|||||||||

{ 299 }

Which number fits in the next spot in this sequence?:

3.5, 7, 14, 28, 56, ...?

||||

By taking the ratios of adjacent terms, we can see that this is a Geometric Progression, with common ratio = 2. Therefore, the sixth term can be calculated by multiplying the fifth term with the common ratio. It is equal to 56 x 2 = 112.

|||||||||

{ 300 }

Which number fits in the next spot in this sequence?:

6.8, 20.4, 61.2, 183.6, 550.8, ...?

||||

By taking the ratios of adjacent terms, we can see that this is a Geometric Progression, with common ratio = 3. Therefore, the sixth term can be calculated by multiplying the fifth term with the common ratio. It is equal to 550.8 x 3 = 1652.4.

|||||||||

{ 301 }

Which number fits in the next spot in this sequence?:

98, 393, 1573, 6293, 25173, ...?

||||

By observation and trial and error, we find that each term in this

sequence is obtained by multiplying the previous term by 4 and adding 1.

Therefore, the sixth term = (4 x 25173) + 1 = 100693.

|||||||||

{ 302 }

Which number fits in the next spot in this sequence?:

191, 574, 1723, 5170, 15511, ...?

||||

By observation and trial and error, we find that each term in this

sequence is obtained by multiplying the previous term by 3 and adding 1.

Therefore, the sixth term = (3 x 15511) + 1 = 46534.

|||||||||

{ 303 }

Which number fits in the next spot in this sequence?:

186, 373, 747, 1495, 2991, ...?

||||

By observation and trial and error, we find that each term in this

sequence is obtained by multiplying the previous term by 2 and adding 1.

Therefore, the sixth term = (2 x 2991) + 1 = 5983.

|||||||||

{ 304 }

Which number fits in the next spot in this sequence?:

7.4, 29.6, 118.4, 473.6, 1894.4, ...?

||||

By taking the ratios of adjacent terms, we can see that this is a

Geometric Progression, with common ratio = 4. Therefore, the sixth term

can be calculated by multiplying the fifth term with the common ratio. It is

equal to 1894.4 x 4 = 7577.6.

|||||||||

{ 305 }

Solve this: what comes next in this sequence?:

109, 144, 179, 214, 249, ...?

||||

By taking the differences between adjacent terms, we can see that this is

an Arithmetic Progression, with common difference = 35. Therefore, the

sixth term can be calculated by adding the common difference to the fifth

term. It is equal to 249 + 35 = 284.

|||||||||

{ 306 }

Which number fits in the next spot in this sequence?:

7.9, 23.7, 71.1, 213.3, 639.900000000001, ...?

||||

By taking the ratios of adjacent terms, we can see that this is a

Geometric Progression, with common ratio = 3. Therefore, the sixth term

can be calculated by multiplying the fifth term with the common ratio. It is

equal to 639.900000000001 x 3 = 1919.7.

|||||||||

{ 307 }

Which number fits in the next spot in this sequence?:

3.9, 15.6, 62.4, 249.6, 998.4, ...?

||||

By taking the ratios of adjacent terms, we can see that this is a

Geometric Progression, with common ratio = 4. Therefore, the sixth term

can be calculated by multiplying the fifth term with the common ratio. It is

equal to 998.4 x 4 = 3993.6.

|||||||||

{ 308 }

Solve this: what comes next in this sequence?:

250, 294, 338, 382, 426, ...?

IIII

By taking the differences between adjacent terms, we can see that this is

an Arithmetic Progression, with common difference = 44. Therefore, the

sixth term can be calculated by adding the common difference to the fifth

term. It is equal to 426 + 44 = 470.

IIIIIIII

{ 309 }

Solve this: what comes next in this sequence?:

93, 114, 135, 156, 177, ...?

IIII

By taking the differences between adjacent terms, we can see that this is

an Arithmetic Progression, with common difference = 21. Therefore, the

sixth term can be calculated by adding the common difference to the fifth

term. It is equal to 177 + 21 = 198.

IIIIIIII

{ 310 }

Solve this: what comes next in this sequence?:

285, 322, 359, 396, 433, ...?

IIII

By taking the differences between adjacent terms, we can see that this is

an Arithmetic Progression, with common difference = 37. Therefore, the

sixth term can be calculated by adding the common difference to the fifth term. It is equal to 433 + 37 = 470.

||||||||

{ 311 }

Solve this: what comes next in this sequence?:

281, 305, 329, 353, 377, ...?

||||

By taking the differences between adjacent terms, we can see that this is an Arithmetic Progression, with common difference = 24. Therefore, the sixth term can be calculated by adding the common difference to the fifth term. It is equal to 377 + 24 = 401.

|||||||||

{ 312 }

Solve this: what comes next in this sequence?:

199, 235, 271, 307, 343, ...?

||||

By taking the differences between adjacent terms, we can see that this is an Arithmetic Progression, with common difference = 36. Therefore, the sixth term can be calculated by adding the common difference to the fifth term. It is equal to 343 + 36 = 379.

|||||||||

{ 313 }

Solve this: what comes next in this sequence?:

100, 121, 142, 163, 184, ...?

IIII

By taking the differences between adjacent terms, we can see that this is an Arithmetic Progression, with common difference = 21. Therefore, the sixth term can be calculated by adding the common difference to the fifth term. It is equal to 184 + 21 = 205.

IIIIIIII

{ 314 }

Solve this: what comes next in this sequence?:

230, 254, 278, 302, 326, ...?

IIII

By taking the differences between adjacent terms, we can see that this is an Arithmetic Progression, with common difference = 24. Therefore, the sixth term can be calculated by adding the common difference to the fifth term. It is equal to 326 + 24 = 350.

IIIIIIII

{ 315 }

Which number fits in the next spot in this sequence?:

142, 285, 571, 1143, 2287, ...?

||||

By observation and trial and error, we find that each term in this sequence is obtained by multiplying the previous term by 2 and adding 1. Therefore, the sixth term = (2 x 2287) + 1 = 4575.

|||||||||

{ 316 }

Solve this: what comes next in this sequence?:

143, 189, 235, 281, 327, ...?

||||

By taking the differences between adjacent terms, we can see that this is an Arithmetic Progression, with common difference = 46. Therefore, the sixth term can be calculated by adding the common difference to the fifth term. It is equal to 327 + 46 = 373.

||||||||

{ 317 }

Which number fits in the next spot in this sequence?:

136, 409, 1228, 3685, 11056, ...?

||||

By observation and trial and error, we find that each term in this sequence is obtained by multiplying the previous term by 3 and adding 1. Therefore, the sixth term = (3 x 11056) + 1 = 33169.

IIIIIIII

{ 318 }

Solve this: what comes next in this sequence?:

106, 119, 132, 145, 158, ...?

IIII

By taking the differences between adjacent terms, we can see that this is an Arithmetic Progression, with common difference = 13. Therefore, the sixth term can be calculated by adding the common difference to the fifth term. It is equal to 158 + 13 = 171.

IIIIIIII

{ 319 }

Which number fits in the next spot in this sequence?:

5.2, 15.6, 46.8, 140.4, 421.2, ...?

IIII

By taking the ratios of adjacent terms, we can see that this is a Geometric Progression, with common ratio = 3. Therefore, the sixth term can be calculated by multiplying the fifth term with the common ratio. It is equal to 421.2 x 3 = 1263.6.

IIIIIIII

{ 320 }

Which number fits in the next spot in this sequence?:

8.8, 26.4, 79.2, 237.6, 712.800000000001, ...?

||||

By taking the ratios of adjacent terms, we can see that this is a Geometric Progression, with common ratio = 3. Therefore, the sixth term can be calculated by multiplying the fifth term with the common ratio. It is equal to 712.800000000001 x 3 = 2138.4.

|||||||||

{ 321 }

Solve this: what comes next in this sequence?:

275, 288, 301, 314, 327, ...?

||||

By taking the differences between adjacent terms, we can see that this is an Arithmetic Progression, with common difference = 13. Therefore, the sixth term can be calculated by adding the common difference to the fifth term. It is equal to 327 + 13 = 340.

||||||||

{ 322 }

Solve this: what comes next in this sequence?:

203, 234, 265, 296, 327, ...?

||||

By taking the differences between adjacent terms, we can see that this is an Arithmetic Progression, with common difference = 31. Therefore, the sixth term can be calculated by adding the common difference to the fifth term. It is equal to 327 + 31 = 358.

|||||||||

{ 323 }

Solve this: what comes next in this sequence?:

108, 152, 196, 240, 284, ...?

||||

By taking the differences between adjacent terms, we can see that this is an Arithmetic Progression, with common difference = 44. Therefore, the sixth term can be calculated by adding the common difference to the fifth term. It is equal to 284 + 44 = 328.

|||||||||

{ 324 }

Which number fits in the next spot in this sequence?:

227, 682, 2047, 6142, 18427, ...?

||||

By observation and trial and error, we find that each term in this sequence is obtained by multiplying the previous term by 3 and adding 1.

Therefore, the sixth term = (3 x 18427) + 1 = 55282.

||||||||

{ 325 }

Which number fits in the next spot in this sequence?:

253, 507, 1015, 2031, 4063, ...?

||||

By observation and trial and error, we find that each term in this

sequence is obtained by multiplying the previous term by 2 and adding 1.

Therefore, the sixth term = (2 x 4063) + 1 = 8127.

||||||||

{ 326 }

Solve this: what comes next in this sequence?:

242, 270, 298, 326, 354, ...?

||||

By taking the differences between adjacent terms, we can see that this is

an Arithmetic Progression, with common difference = 28. Therefore, the

sixth term can be calculated by adding the common difference to the fifth

term. It is equal to 354 + 28 = 382.

||||||||

{ 327 }

Which number fits in the next spot in this sequence?:

9.1, 18.2, 36.4, 72.8, 145.6, ...?

||||

By taking the ratios of adjacent terms, we can see that this is a

Geometric Progression, with common ratio = 2. Therefore, the sixth term

can be calculated by multiplying the fifth term with the common ratio. It is

equal to 145.6 x 2 = 291.2.

|||||||||

{ 328 }

Solve this: what comes next in this sequence?:

187, 209, 231, 253, 275, ...?

||||

By taking the differences between adjacent terms, we can see that this is

an Arithmetic Progression, with common difference = 22. Therefore, the

sixth term can be calculated by adding the common difference to the fifth

term. It is equal to 275 + 22 = 297.

|||||||||

{ 329 }

Which number fits in the next spot in this sequence?:

1.1, 3.3, 9.9, 29.7, 89.1000000000001, ...?

||||

By taking the ratios of adjacent terms, we can see that this is a

Geometric Progression, with common ratio = 3. Therefore, the sixth term

can be calculated by multiplying the fifth term with the common ratio. It is

equal to 89.1000000000001 x 3 = 267.3.

|||||||||

{ 330 }

Which number fits in the next spot in this sequence?:

8.2, 24.6, 73.8, 221.4, 664.2, ...?

||||

By taking the ratios of adjacent terms, we can see that this is a

Geometric Progression, with common ratio = 3. Therefore, the sixth term

can be calculated by multiplying the fifth term with the common ratio. It is

equal to 664.2 x 3 = 1992.6.

|||||||||

{ 331 }

Which number fits in the next spot in this sequence?:

9.1, 27.3, 81.9, 245.7, 737.1, ...?

||||

By taking the ratios of adjacent terms, we can see that this is a

Geometric Progression, with common ratio = 3. Therefore, the sixth term

can be calculated by multiplying the fifth term with the common ratio. It is equal to 737.1 x 3 = 2211.3.

|||||||||

{ 332 }

Which number fits in the next spot in this sequence?:

7, 21, 63, 189, 567, ...?

||||

By taking the ratios of adjacent terms, we can see that this is a Geometric Progression, with common ratio = 3. Therefore, the sixth term can be calculated by multiplying the fifth term with the common ratio. It is equal to 567 x 3 = 1701.

|||||||||

{ 333 }

Which number fits in the next spot in this sequence?:

7.3, 29.2, 116.8, 467.2, 1868.8, ...?

||||

By taking the ratios of adjacent terms, we can see that this is a Geometric Progression, with common ratio = 4. Therefore, the sixth term can be calculated by multiplying the fifth term with the common ratio. It is equal to 1868.8 x 4 = 7475.2.

|||||||||

{ 334 }

Which number fits in the next spot in this sequence?:

210, 421, 843, 1687, 3375, ...?

||||

By observation and trial and error, we find that each term in this

sequence is obtained by multiplying the previous term by 2 and adding 1.

Therefore, the sixth term = (2 x 3375) + 1 = 6751.

||||||||

{ 335 }

Which number fits in the next spot in this sequence?:

190, 381, 763, 1527, 3055, ...?

||||

By observation and trial and error, we find that each term in this

sequence is obtained by multiplying the previous term by 2 and adding 1.

Therefore, the sixth term = (2 x 3055) + 1 = 6111.

||||||||

{ 336 }

Solve this: what comes next in this sequence?:

262, 282, 302, 322, 342, ...?

||||

By taking the differences between adjacent terms, we can see that this is

an Arithmetic Progression, with common difference = 20. Therefore, the sixth term can be calculated by adding the common difference to the fifth term. It is equal to 342 + 20 = 362.

||||||||

{ 337 }

Which number fits in the next spot in this sequence?:

2.9, 8.7, 26.1, 78.3, 234.9, ...?

||||

By taking the ratios of adjacent terms, we can see that this is a Geometric Progression, with common ratio = 3. Therefore, the sixth term can be calculated by multiplying the fifth term with the common ratio. It is equal to 234.9 x 3 = 704.7.

||||||||

{ 338 }

Which number fits in the next spot in this sequence?:

8.7, 17.4, 34.8, 69.6, 139.2, ...?

||||

By taking the ratios of adjacent terms, we can see that this is a Geometric Progression, with common ratio = 2. Therefore, the sixth term can be calculated by multiplying the fifth term with the common ratio. It is equal to 139.2 x 2 = 278.4.

IIIIIIII

{ 339 }

Which number fits in the next spot in this sequence?:

3.3, 13.2, 52.8, 211.2, 844.8, ...?

IIII

By taking the ratios of adjacent terms, we can see that this is a

Geometric Progression, with common ratio = 4. Therefore, the sixth term

can be calculated by multiplying the fifth term with the common ratio. It is

equal to 844.8 x 4 = 3379.2.

IIIIIIII

{ 340 }

Which number fits in the next spot in this sequence?:

147, 589, 2357, 9429, 37717, ...?

IIII

By observation and trial and error, we find that each term in this

sequence is obtained by multiplying the previous term by 4 and adding 1.

Therefore, the sixth term = (4 x 37717) + 1 = 150869.

IIIIIIII

{ 341 }

Solve this: what comes next in this sequence?:

150, 200, 250, 300, 350, ...?

||||

By taking the differences between adjacent terms, we can see that this is an Arithmetic Progression, with common difference = 50. Therefore, the sixth term can be calculated by adding the common difference to the fifth term. It is equal to 350 + 50 = 400.

|||||||||

{ 342 }

Which number fits in the next spot in this sequence?:

2.7, 10.8, 43.2, 172.8, 691.2, ...?

||||

By taking the ratios of adjacent terms, we can see that this is a Geometric Progression, with common ratio = 4. Therefore, the sixth term can be calculated by multiplying the fifth term with the common ratio. It is equal to 691.2 x 4 = 2764.8.

|||||||||

{ 343 }

Which number fits in the next spot in this sequence?:

3.1, 9.3, 27.9, 83.7, 251.1, ...?

||||

By taking the ratios of adjacent terms, we can see that this is a

Geometric Progression, with common ratio = 3. Therefore, the sixth term

can be calculated by multiplying the fifth term with the common ratio. It is

equal to 251.1 x 3 = 753.3.

|||||||||

{ 344 }

Which number fits in the next spot in this sequence?:

3.8, 7.6, 15.2, 30.4, 60.8, ...?

||||

By taking the ratios of adjacent terms, we can see that this is a

Geometric Progression, with common ratio = 2. Therefore, the sixth term

can be calculated by multiplying the fifth term with the common ratio. It is

equal to 60.8 x 2 = 121.6.

|||||||||

{ 345 }

Solve this: what comes next in this sequence?:

226, 240, 254, 268, 282, ...?

||||

By taking the differences between adjacent terms, we can see that this is

an Arithmetic Progression, with common difference = 14. Therefore, the

sixth term can be calculated by adding the common difference to the fifth

term. It is equal to 282 + 14 = 296.

IIIIIIII

{ 346 }

Which number fits in the next spot in this sequence?:

213, 427, 855, 1711, 3423, ...?

IIII

By observation and trial and error, we find that each term in this

sequence is obtained by multiplying the previous term by 2 and adding 1.

Therefore, the sixth term = (2 x 3423) + 1 = 6847.

IIIIIIII

{ 347 }

Which number fits in the next spot in this sequence?:

8.9, 26.7, 80.1, 240.3, 720.9, ...?

IIII

By taking the ratios of adjacent terms, we can see that this is a

Geometric Progression, with common ratio = 3. Therefore, the sixth term

can be calculated by multiplying the fifth term with the common ratio. It is

equal to 720.9 x 3 = 2162.7.

IIIIIIII

{ 348 }

Which number fits in the next spot in this sequence?:

1.2, 3.6, 10.8, 32.4, 97.2, ...?

||||

By taking the ratios of adjacent terms, we can see that this is a Geometric Progression, with common ratio = 3. Therefore, the sixth term can be calculated by multiplying the fifth term with the common ratio. It is equal to 97.2 x 3 = 291.6.

|||||||||

{ 349 }

Solve this: what comes next in this sequence?:

284, 295, 306, 317, 328, ...?

||||

By taking the differences between adjacent terms, we can see that this is an Arithmetic Progression, with common difference = 11. Therefore, the sixth term can be calculated by adding the common difference to the fifth term. It is equal to 328 + 11 = 339.

|||||||||

{ 350 }

Solve this: what comes next in this sequence?:

122, 142, 162, 182, 202, ...?

||||

By taking the differences between adjacent terms, we can see that this is

an Arithmetic Progression, with common difference = 20. Therefore, the

sixth term can be calculated by adding the common difference to the fifth

term. It is equal to 202 + 20 = 222.

|||||||||

{ 351 }

Which number fits in the next spot in this sequence?:

230, 461, 923, 1847, 3695, ...?

||||

By observation and trial and error, we find that each term in this

sequence is obtained by multiplying the previous term by 2 and adding 1.

Therefore, the sixth term = (2 x 3695) + 1 = 7391.

|||||||||

{ 352 }

Which number fits in the next spot in this sequence?:

9.6, 19.2, 38.4, 76.8, 153.6, ...?

||||

By taking the ratios of adjacent terms, we can see that this is a

Geometric Progression, with common ratio = 2. Therefore, the sixth term

can be calculated by multiplying the fifth term with the common ratio. It is

equal to 153.6 x 2 = 307.2.

||||||||

{ 353 }

Which number fits in the next spot in this sequence?:

153, 307, 615, 1231, 2463, ...?

||||

By observation and trial and error, we find that each term in this

sequence is obtained by multiplying the previous term by 2 and adding 1.

Therefore, the sixth term = (2 x 2463) + 1 = 4927.

||||||||

{ 354 }

Solve this: what comes next in this sequence?:

170, 216, 262, 308, 354, ...?

||||

By taking the differences between adjacent terms, we can see that this is

an Arithmetic Progression, with common difference = 46. Therefore, the

sixth term can be calculated by adding the common difference to the fifth

term. It is equal to 354 + 46 = 400.

||||||||

{ 355 }

Solve this: what comes next in this sequence?:

212, 262, 312, 362, 412, ...?

||||

By taking the differences between adjacent terms, we can see that this is

an Arithmetic Progression, with common difference = 50. Therefore, the

sixth term can be calculated by adding the common difference to the fifth

term. It is equal to 412 + 50 = 462.

|||||||||

{ 356 }

Which number fits in the next spot in this sequence?:

3.9, 7.8, 15.6, 31.2, 62.4, ...?

||||

By taking the ratios of adjacent terms, we can see that this is a

Geometric Progression, with common ratio = 2. Therefore, the sixth term

can be calculated by multiplying the fifth term with the common ratio. It is

equal to 62.4 x 2 = 124.8.

|||||||||

{ 357 }

Which number fits in the next spot in this sequence?:

7, 14, 28, 56, 112, ...?

||||

By taking the ratios of adjacent terms, we can see that this is a

Geometric Progression, with common ratio = 2. Therefore, the sixth term

can be calculated by multiplying the fifth term with the common ratio. It is

equal to 112 x 2 = 224.

|||||||||

{ 358 }

Solve this: what comes next in this sequence?:

126, 150, 174, 198, 222, ...?

||||

By taking the differences between adjacent terms, we can see that this is

an Arithmetic Progression, with common difference = 24. Therefore, the

sixth term can be calculated by adding the common difference to the fifth

term. It is equal to 222 + 24 = 246.

|||||||||

{ 359 }

Which number fits in the next spot in this sequence?:

170, 341, 683, 1367, 2735, ...?

||||

By observation and trial and error, we find that each term in this

sequence is obtained by multiplying the previous term by 2 and adding 1.

Therefore, the sixth term = (2 x 2735) + 1 = 5471.

IIIIIIIII

{ 360 }

Which number fits in the next spot in this sequence?:

127, 509, 2037, 8149, 32597, ...?

IIII

By observation and trial and error, we find that each term in this

sequence is obtained by multiplying the previous term by 4 and adding 1.

Therefore, the sixth term = (4 x 32597) + 1 = 130389.

IIIIIIIII

{ 361 }

Solve this: what comes next in this sequence?:

126, 168, 210, 252, 294, ...?

IIII

By taking the differences between adjacent terms, we can see that this is

an Arithmetic Progression, with common difference = 42. Therefore, the

sixth term can be calculated by adding the common difference to the fifth

term. It is equal to 294 + 42 = 336.

IIIIIIIII

{ 362 }

Which number fits in the next spot in this sequence?:

204, 613, 1840, 5521, 16564, ...?

||||

By observation and trial and error, we find that each term in this sequence is obtained by multiplying the previous term by 3 and adding 1. Therefore, the sixth term = (3 x 16564) + 1 = 49693.

|||||||||

{ 363 }

Which number fits in the next spot in this sequence?:

2.5, 10, 40, 160, 640, ...?

||||

By taking the ratios of adjacent terms, we can see that this is a Geometric Progression, with common ratio = 4. Therefore, the sixth term can be calculated by multiplying the fifth term with the common ratio. It is equal to 640 x 4 = 2560.

|||||||||

{ 364 }

Which number fits in the next spot in this sequence?:

234, 469, 939, 1879, 3759, ...?

||||

By observation and trial and error, we find that each term in this

sequence is obtained by multiplying the previous term by 2 and adding 1.

Therefore, the sixth term = (2 x 3759) + 1 = 7519.

|||||||||

{ 365 }

Which number fits in the next spot in this sequence?:

155, 466, 1399, 4198, 12595, ...?

||||

By observation and trial and error, we find that each term in this

sequence is obtained by multiplying the previous term by 3 and adding 1.

Therefore, the sixth term = (3 x 12595) + 1 = 37786.

|||||||||

{ 366 }

Which number fits in the next spot in this sequence?:

2.7, 10.8, 43.2, 172.8, 691.2, ...?

||||

By taking the ratios of adjacent terms, we can see that this is a

Geometric Progression, with common ratio = 4. Therefore, the sixth term

can be calculated by multiplying the fifth term with the common ratio. It is

equal to 691.2 x 4 = 2764.8.

|||||||||

{ 367 }

Which number fits in the next spot in this sequence?:

138, 553, 2213, 8853, 35413, ...?

||||

By observation and trial and error, we find that each term in this sequence is obtained by multiplying the previous term by 4 and adding 1. Therefore, the sixth term = (4 x 35413) + 1 = 141653.

|||||||||

{ 368 }

Solve this: what comes next in this sequence?:

96, 112, 128, 144, 160, ...?

||||

By taking the differences between adjacent terms, we can see that this is an Arithmetic Progression, with common difference = 16. Therefore, the sixth term can be calculated by adding the common difference to the fifth term. It is equal to 160 + 16 = 176.

|||||||||

{ 369 }

Solve this: what comes next in this sequence?:

117, 127, 137, 147, 157, ...?

||||

By taking the differences between adjacent terms, we can see that this is an Arithmetic Progression, with common difference = 10. Therefore, the sixth term can be calculated by adding the common difference to the fifth term. It is equal to 157 + 10 = 167.

|||||||||

{ 370 }

Which number fits in the next spot in this sequence?:

4.8, 14.4, 43.2, 129.6, 388.8, ...?

|||||

By taking the ratios of adjacent terms, we can see that this is a Geometric Progression, with common ratio = 3. Therefore, the sixth term can be calculated by multiplying the fifth term with the common ratio. It is equal to 388.8 x 3 = 1166.4.

|||||||||

{ 371 }

Which number fits in the next spot in this sequence?:

1.9, 5.7, 17.1, 51.3, 153.9, ...?

|||||

By taking the ratios of adjacent terms, we can see that this is a Geometric Progression, with common ratio = 3. Therefore, the sixth term can be calculated by multiplying the fifth term with the common ratio. It is

equal to 153.9 x 3 = 461.7.

||||||||

{ 372 }

Which number fits in the next spot in this sequence?:

158, 317, 635, 1271, 2543, ...?

||||

By observation and trial and error, we find that each term in this

sequence is obtained by multiplying the previous term by 2 and adding 1.

Therefore, the sixth term = (2 x 2543) + 1 = 5087.

||||||||

{ 373 }

Which number fits in the next spot in this sequence?:

8.2, 32.8, 131.2, 524.8, 2099.2, ...?

||||

By taking the ratios of adjacent terms, we can see that this is a

Geometric Progression, with common ratio = 4. Therefore, the sixth term

can be calculated by multiplying the fifth term with the common ratio. It is

equal to 2099.2 x 4 = 8396.8.

||||||||

{ 374 }

Which number fits in the next spot in this sequence?:

8.1, 24.3, 72.9, 218.7, 656.1, ...?

||||

By taking the ratios of adjacent terms, we can see that this is a

Geometric Progression, with common ratio = 3. Therefore, the sixth term

can be calculated by multiplying the fifth term with the common ratio. It is

equal to 656.1 x 3 = 1968.3.

|||||||||

{ 375 }

Which number fits in the next spot in this sequence?:

10, 20, 40, 80, 160, ...?

||||

By taking the ratios of adjacent terms, we can see that this is a

Geometric Progression, with common ratio = 2. Therefore, the sixth term

can be calculated by multiplying the fifth term with the common ratio. It is

equal to 160 x 2 = 320.

|||||||||

{ 376 }

Which number fits in the next spot in this sequence?:

274, 1097, 4389, 17557, 70229, ...?

||||

By observation and trial and error, we find that each term in this sequence is obtained by multiplying the previous term by 4 and adding 1. Therefore, the sixth term = (4 x 70229) + 1 = 280917.

|||||||||

{ 377 }

Solve this: what comes next in this sequence?:

165, 184, 203, 222, 241, ...?

||||

By taking the differences between adjacent terms, we can see that this is an Arithmetic Progression, with common difference = 19. Therefore, the sixth term can be calculated by adding the common difference to the fifth term. It is equal to 241 + 19 = 260.

|||||||||

{ 378 }

Solve this: what comes next in this sequence?:

196, 222, 248, 274, 300, ...?

||||

By taking the differences between adjacent terms, we can see that this is an Arithmetic Progression, with common difference = 26. Therefore, the sixth term can be calculated by adding the common difference to the fifth

term. It is equal to 300 + 26 = 326.

||||||||

{ 379 }

Which number fits in the next spot in this sequence?:

93, 280, 841, 2524, 7573, ...?

||||

By observation and trial and error, we find that each term in this

sequence is obtained by multiplying the previous term by 3 and adding 1.

Therefore, the sixth term = (3 x 7573) + 1 = 22720.

||||||||

{ 380 }

Which number fits in the next spot in this sequence?:

2.1, 4.2, 8.4, 16.8, 33.6, ...?

||||

By taking the ratios of adjacent terms, we can see that this is a

Geometric Progression, with common ratio = 2. Therefore, the sixth term

can be calculated by multiplying the fifth term with the common ratio. It is

equal to 33.6 x 2 = 67.2.

||||||||

{ 381 }

Solve this: what comes next in this sequence?:

291, 323, 355, 387, 419, ...?

IIII

By taking the differences between adjacent terms, we can see that this is

an Arithmetic Progression, with common difference = 32. Therefore, the

sixth term can be calculated by adding the common difference to the fifth

term. It is equal to 419 + 32 = 451.

IIIIIIII

{ 382 }

Solve this: what comes next in this sequence?:

176, 207, 238, 269, 300, ...?

IIII

By taking the differences between adjacent terms, we can see that this is

an Arithmetic Progression, with common difference = 31. Therefore, the

sixth term can be calculated by adding the common difference to the fifth

term. It is equal to 300 + 31 = 331.

IIIIIIII

{ 383 }

Solve this: what comes next in this sequence?:

258, 273, 288, 303, 318, ...?

||||

By taking the differences between adjacent terms, we can see that this is an Arithmetic Progression, with common difference = 15. Therefore, the sixth term can be calculated by adding the common difference to the fifth term. It is equal to 318 + 15 = 333.

|||||||||

{ 384 }

Which number fits in the next spot in this sequence?:

6.1, 12.2, 24.4, 48.8, 97.6, ...?

||||

By taking the ratios of adjacent terms, we can see that this is a Geometric Progression, with common ratio = 2. Therefore, the sixth term can be calculated by multiplying the fifth term with the common ratio. It is equal to 97.6 x 2 = 195.2.

|||||||||

{ 385 }

Which number fits in the next spot in this sequence?:

6.7, 13.4, 26.8, 53.6, 107.2, ...?

||||

By taking the ratios of adjacent terms, we can see that this is a Geometric Progression, with common ratio = 2. Therefore, the sixth term

can be calculated by multiplying the fifth term with the common ratio. It is equal to 107.2 x 2 = 214.4.

||||||||

{ 386 }

Which number fits in the next spot in this sequence?:

138, 277, 555, 1111, 2223, ...?

||||

By observation and trial and error, we find that each term in this sequence is obtained by multiplying the previous term by 2 and adding 1. Therefore, the sixth term = (2 x 2223) + 1 = 4447.

|||||||||

{ 387 }

Which number fits in the next spot in this sequence?:

169, 677, 2709, 10837, 43349, ...?

||||

By observation and trial and error, we find that each term in this sequence is obtained by multiplying the previous term by 4 and adding 1. Therefore, the sixth term = (4 x 43349) + 1 = 173397.

|||||||||

{ 388 }

Which number fits in the next spot in this sequence?:

153, 613, 2453, 9813, 39253, ...?

||||

By observation and trial and error, we find that each term in this
sequence is obtained by multiplying the previous term by 4 and adding 1.
Therefore, the sixth term = (4 x 39253) + 1 = 157013.

|||||||||

{ 389 }

Which number fits in the next spot in this sequence?:

3, 9, 27, 81, 243, ...?

||||

By taking the ratios of adjacent terms, we can see that this is a
Geometric Progression, with common ratio = 3. Therefore, the sixth term
can be calculated by multiplying the fifth term with the common ratio. It is
equal to 243 x 3 = 729.

|||||||||

{ 390 }

Which number fits in the next spot in this sequence?:

3.4, 6.8, 13.6, 27.2, 54.4, ...?

||||

By taking the ratios of adjacent terms, we can see that this is a Geometric Progression, with common ratio = 2. Therefore, the sixth term can be calculated by multiplying the fifth term with the common ratio. It is equal to 54.4 x 2 = 108.8.

|||||||||

{ 391 }

Which number fits in the next spot in this sequence?:

225, 676, 2029, 6088, 18265, ...?

|||||

By observation and trial and error, we find that each term in this sequence is obtained by multiplying the previous term by 3 and adding 1. Therefore, the sixth term = (3 x 18265) + 1 = 54796.

|||||||||

{ 392 }

Which number fits in the next spot in this sequence?:

4.8, 19.2, 76.8, 307.2, 1228.8, ...?

|||||

By taking the ratios of adjacent terms, we can see that this is a Geometric Progression, with common ratio = 4. Therefore, the sixth term can be calculated by multiplying the fifth term with the common ratio. It is equal to 1228.8 x 4 = 4915.2.

|||||||||

{ 393 }

Solve this: what comes next in this sequence?:

295, 344, 393, 442, 491, ...?

||||

By taking the differences between adjacent terms, we can see that this is an Arithmetic Progression, with common difference = 49. Therefore, the sixth term can be calculated by adding the common difference to the fifth term. It is equal to 491 + 49 = 540.

|||||||||

{ 394 }

Solve this: what comes next in this sequence?:

165, 181, 197, 213, 229, ...?

||||

By taking the differences between adjacent terms, we can see that this is an Arithmetic Progression, with common difference = 16. Therefore, the sixth term can be calculated by adding the common difference to the fifth term. It is equal to 229 + 16 = 245.

|||||||||

{ 395 }

Solve this: what comes next in this sequence?:

256, 296, 336, 376, 416, ...?

IIII

By taking the differences between adjacent terms, we can see that this is

an Arithmetic Progression, with common difference = 40. Therefore, the

sixth term can be calculated by adding the common difference to the fifth

term. It is equal to 416 + 40 = 456.

IIIIIIII

{ 396 }

Which number fits in the next spot in this sequence?:

137, 549, 2197, 8789, 35157, ...?

IIII

By observation and trial and error, we find that each term in this

sequence is obtained by multiplying the previous term by 4 and adding 1.

Therefore, the sixth term = (4 x 35157) + 1 = 140629.

IIIIIIII

{ 397 }

Which number fits in the next spot in this sequence?:

7.1, 14.2, 28.4, 56.8, 113.6, ...?

IIII

By taking the ratios of adjacent terms, we can see that this is a Geometric Progression, with common ratio = 2. Therefore, the sixth term can be calculated by multiplying the fifth term with the common ratio. It is equal to 113.6 x 2 = 227.2.

IIIIIIII

{ 398 }

Solve this: what comes next in this sequence?:

234, 276, 318, 360, 402, ...?

IIII

By taking the differences between adjacent terms, we can see that this is an Arithmetic Progression, with common difference = 42. Therefore, the sixth term can be calculated by adding the common difference to the fifth term. It is equal to 402 + 42 = 444.

IIIIIIII

{ 399 }

Solve this: what comes next in this sequence?:

251, 280, 309, 338, 367, ...?

IIII

By taking the differences between adjacent terms, we can see that this is an Arithmetic Progression, with common difference = 29. Therefore, the sixth term can be calculated by adding the common difference to the fifth

term. It is equal to 367 + 29 = 396.

||||||||

{ 400 }

Solve this: what comes next in this sequence?:

244, 286, 328, 370, 412, ...?

||||

By taking the differences between adjacent terms, we can see that this is

an Arithmetic Progression, with common difference = 42. Therefore, the

sixth term can be calculated by adding the common difference to the fifth

term. It is equal to 412 + 42 = 454.

||||||||

{ 401 }

Which number fits in the next spot in this sequence?:

5, 15, 45, 135, 405, ...?

||||

By taking the ratios of adjacent terms, we can see that this is a

Geometric Progression, with common ratio = 3. Therefore, the sixth term

can be calculated by multiplying the fifth term with the common ratio. It is

equal to 405 x 3 = 1215.

||||||||

{ 402 }

Which number fits in the next spot in this sequence?:

119, 358, 1075, 3226, 9679, ...?

IIII

By observation and trial and error, we find that each term in this

sequence is obtained by multiplying the previous term by 3 and adding 1.

Therefore, the sixth term = (3 x 9679) + 1 = 29038.

IIIIIIII

{ 403 }

Which number fits in the next spot in this sequence?:

9.5, 28.5, 85.5, 256.5, 769.5, ...?

IIII

By taking the ratios of adjacent terms, we can see that this is a

Geometric Progression, with common ratio = 3. Therefore, the sixth term

can be calculated by multiplying the fifth term with the common ratio. It is

equal to 769.5 x 3 = 2308.5.

IIIIIIII

{ 404 }

Which number fits in the next spot in this sequence?:

290, 581, 1163, 2327, 4655, ...?

IIII

By observation and trial and error, we find that each term in this sequence is obtained by multiplying the previous term by 2 and adding 1.

Therefore, the sixth term = (2 x 4655) + 1 = 9311.

||||||||

{ 405 }

Which number fits in the next spot in this sequence?:

198, 595, 1786, 5359, 16078, ...?

||||

By observation and trial and error, we find that each term in this sequence is obtained by multiplying the previous term by 3 and adding 1.

Therefore, the sixth term = (3 x 16078) + 1 = 48235.

|||||||||

{ 406 }

Which number fits in the next spot in this sequence?:

232, 697, 2092, 6277, 18832, ...?

||||

By observation and trial and error, we find that each term in this sequence is obtained by multiplying the previous term by 3 and adding 1.

Therefore, the sixth term = (3 x 18832) + 1 = 56497.

|||||||||

{ 407 }

Which number fits in the next spot in this sequence?:

8.5, 34, 136, 544, 2176, ...?

||||

By taking the ratios of adjacent terms, we can see that this is a

Geometric Progression, with common ratio = 4. Therefore, the sixth term

can be calculated by multiplying the fifth term with the common ratio. It is

equal to 2176 x 4 = 8704.

|||||||||

{ 408 }

Which number fits in the next spot in this sequence?:

5.2, 15.6, 46.8, 140.4, 421.2, ...?

||||

By taking the ratios of adjacent terms, we can see that this is a

Geometric Progression, with common ratio = 3. Therefore, the sixth term

can be calculated by multiplying the fifth term with the common ratio. It is

equal to 421.2 x 3 = 1263.6.

|||||||||

{ 409 }

Which number fits in the next spot in this sequence?:

288, 1153, 4613, 18453, 73813, ...?

||||

By observation and trial and error, we find that each term in this sequence is obtained by multiplying the previous term by 4 and adding 1. Therefore, the sixth term = (4 x 73813) + 1 = 295253.

|||||||||

{ 410 }

Which number fits in the next spot in this sequence?:

6.5, 19.5, 58.5, 175.5, 526.5, ...?

||||

By taking the ratios of adjacent terms, we can see that this is a Geometric Progression, with common ratio = 3. Therefore, the sixth term can be calculated by multiplying the fifth term with the common ratio. It is equal to 526.5 x 3 = 1579.5.

||||||||

{ 411 }

Solve this: what comes next in this sequence?:

94, 117, 140, 163, 186, ...?

||||

By taking the differences between adjacent terms, we can see that this is an Arithmetic Progression, with common difference = 23. Therefore, the sixth term can be calculated by adding the common difference to the fifth

term. It is equal to 186 + 23 = 209.

||||||||

{ 412 }

Which number fits in the next spot in this sequence?:

231, 463, 927, 1855, 3711, ...?

||||

By observation and trial and error, we find that each term in this

sequence is obtained by multiplying the previous term by 2 and adding 1.

Therefore, the sixth term = (2 x 3711) + 1 = 7423.

||||||||

{ 413 }

Which number fits in the next spot in this sequence?:

9.3, 37.2, 148.8, 595.2, 2380.8, ...?

||||

By taking the ratios of adjacent terms, we can see that this is a

Geometric Progression, with common ratio = 4. Therefore, the sixth term

can be calculated by multiplying the fifth term with the common ratio. It is

equal to 2380.8 x 4 = 9523.2.

||||||||

{ 414 }

Which number fits in the next spot in this sequence?:

238, 953, 3813, 15253, 61013, ...?

||||

By observation and trial and error, we find that each term in this

sequence is obtained by multiplying the previous term by 4 and adding 1.

Therefore, the sixth term = (4 x 61013) + 1 = 244053.

|||||||||

{ 415 }

Which number fits in the next spot in this sequence?:

1.2, 2.4, 4.8, 9.6, 19.2, ...?

||||

By taking the ratios of adjacent terms, we can see that this is a

Geometric Progression, with common ratio = 2. Therefore, the sixth term

can be calculated by multiplying the fifth term with the common ratio. It is

equal to 19.2 x 2 = 38.4.

|||||||||

{ 416 }

Which number fits in the next spot in this sequence?:

173, 520, 1561, 4684, 14053, ...?

||||

By observation and trial and error, we find that each term in this sequence is obtained by multiplying the previous term by 3 and adding 1. Therefore, the sixth term = (3 x 14053) + 1 = 42160.

|||||||||

{ 417 }

Solve this: what comes next in this sequence?:

210, 231, 252, 273, 294, ...?

||||

By taking the differences between adjacent terms, we can see that this is an Arithmetic Progression, with common difference = 21. Therefore, the sixth term can be calculated by adding the common difference to the fifth term. It is equal to 294 + 21 = 315.

|||||||||

{ 418 }

Which number fits in the next spot in this sequence?:

242, 727, 2182, 6547, 19642, ...?

||||

By observation and trial and error, we find that each term in this sequence is obtained by multiplying the previous term by 3 and adding 1. Therefore, the sixth term = (3 x 19642) + 1 = 58927.

|||||||||

{ 419 }

Solve this: what comes next in this sequence?:

124, 147, 170, 193, 216, ...?

||||

By taking the differences between adjacent terms, we can see that this is

an Arithmetic Progression, with common difference = 23. Therefore, the

sixth term can be calculated by adding the common difference to the fifth

term. It is equal to 216 + 23 = 239.

|||||||||

{ 420 }

Which number fits in the next spot in this sequence?:

118, 473, 1893, 7573, 30293, ...?

||||

By observation and trial and error, we find that each term in this

sequence is obtained by multiplying the previous term by 4 and adding 1.

Therefore, the sixth term = (4 x 30293) + 1 = 121173.

|||||||||

Series-ly: Part 4

{ 421 }

Solve this: what comes next in this sequence?:

259, 308, 357, 406, 455, ...?

||||

{ 422 }

Solve this: what comes next in this sequence?:

270, 319, 368, 417, 466, ...?

||||

{ 423 }

Which number fits in the next spot in this sequence?:

6.4, 19.2, 57.6, 172.8, 518.4, ...?

||||

{ 424 }

Which number fits in the next spot in this sequence?:

299, 599, 1199, 2399, 4799, ...?

||||

{ 425 }

Which number fits in the next spot in this sequence?:

7.1, 21.3, 63.9, 191.7, 575.1, ...?

||||

{ 426 }

Which number fits in the next spot in this sequence?:

3.5, 10.5, 31.5, 94.5, 283.5, ...?

||||

{ 427 }

Solve this: what comes next in this sequence?:

150, 194, 238, 282, 326, ...?

||||

{ 428 }

Solve this: what comes next in this sequence?:

217, 252, 287, 322, 357, ...?

||||

{ 429 }

Which number fits in the next spot in this sequence?:

92, 185, 371, 743, 1487, ...?

||||

{ 430 }

Which number fits in the next spot in this sequence?:

142, 427, 1282, 3847, 11542, ...?

||||

{ 431 }

Which number fits in the next spot in this sequence?:

6, 12, 24, 48, 96, ...?

||||

{ 432 }

Solve this: what comes next in this sequence?:

217, 260, 303, 346, 389, ...?

||||

{ 433 }

Solve this: what comes next in this sequence?:

221, 237, 253, 269, 285, ...?

||||

{ 434 }

Solve this: what comes next in this sequence?:

216, 261, 306, 351, 396, ...?

||||

{ 435 }

Which number fits in the next spot in this sequence?:

142, 569, 2277, 9109, 36437, ...?

||||

{ 436 }

Which number fits in the next spot in this sequence?:

93, 187, 375, 751, 1503, ...?

||||

{ 437 }

Solve this: what comes next in this sequence?:

276, 312, 348, 384, 420, ...?

||||

{ 438 }

Solve this: what comes next in this sequence?:

94, 122, 150, 178, 206, ...?

||||

{ 439 }

Which number fits in the next spot in this sequence?:

153, 613, 2453, 9813, 39253, ...?

||||

{ 440 }

Which number fits in the next spot in this sequence?:

268, 537, 1075, 2151, 4303, ...?

||||

{ 441 }

Which number fits in the next spot in this sequence?:

2.8, 8.4, 25.2, 75.6, 226.8, ...?

||||

{ 442 }

Which number fits in the next spot in this sequence?:

7.7, 15.4, 30.8, 61.6, 123.2, ...?

||||

{ 443 }

Solve this: what comes next in this sequence?:

169, 208, 247, 286, 325, ...?

||||

{ 444 }

Which number fits in the next spot in this sequence?:

164, 657, 2629, 10517, 42069, ...?

||||

{ 445 }

Which number fits in the next spot in this sequence?:

3.5, 14, 56, 224, 896, ...?

||||

{ 446 }

Solve this: what comes next in this sequence?:

177, 197, 217, 237, 257, ...?

||||

{ 447 }

Which number fits in the next spot in this sequence?:

214, 857, 3429, 13717, 54869, ...?

||||

{ 448 }

Which number fits in the next spot in this sequence?:

141, 565, 2261, 9045, 36181, ...?

||||

{ 449 }

Which number fits in the next spot in this sequence?:

274, 549, 1099, 2199, 4399, ...?

||||

{ 450 }

Which number fits in the next spot in this sequence?:

4.7, 9.4, 18.8, 37.6, 75.2, ...?

||||

{ 451 }

Solve this: what comes next in this sequence?:

231, 275, 319, 363, 407, ...?

||||

{ 452 }

Which number fits in the next spot in this sequence?:

4.9, 19.6, 78.4, 313.6, 1254.4, ...?

||||

{ 453 }

Which number fits in the next spot in this sequence?:

182, 547, 1642, 4927, 14782, ...?

||||

{ 454 }

Solve this: what comes next in this sequence?:

249, 290, 331, 372, 413, ...?

||||

{ 455 }

Solve this: what comes next in this sequence?:

111, 147, 183, 219, 255, ...?

||||

{ 456 }

Which number fits in the next spot in this sequence?:

6, 18, 54, 162, 486, ...?

||||

{ 457 }

Which number fits in the next spot in this sequence?:

5.2, 10.4, 20.8, 41.6, 83.2, ...?

||||

{ 458 }

Solve this: what comes next in this sequence?:

291, 314, 337, 360, 383, ...?

||||

{ 459 }

Which number fits in the next spot in this sequence?:

3, 9, 27, 81, 243, ...?

||||

{ 460 }

Which number fits in the next spot in this sequence?:

204, 409, 819, 1639, 3279, ...?

||||

{ 461 }

Which number fits in the next spot in this sequence?:

5, 20, 80, 320, 1280, ...?

||||

{ 462 }

Which number fits in the next spot in this sequence?:

165, 661, 2645, 10581, 42325, ...?

||||

{ 463 }

Which number fits in the next spot in this sequence?:

188, 565, 1696, 5089, 15268, ...?

||||

{ 464 }

Which number fits in the next spot in this sequence?:

2.9, 8.7, 26.1, 78.3, 234.9, ...?

||||

{ 465 }

Which number fits in the next spot in this sequence?:

284, 1137, 4549, 18197, 72789, ...?

||||

{ 466 }

Which number fits in the next spot in this sequence?:

160, 481, 1444, 4333, 13000, ...?

||||

{ 467 }

Which number fits in the next spot in this sequence?:

7.5, 22.5, 67.5, 202.5, 607.5, ...?

||||

{ 468 }

Which number fits in the next spot in this sequence?:

266, 1065, 4261, 17045, 68181, ...?

||||

{ 469 }

Which number fits in the next spot in this sequence?:

267, 802, 2407, 7222, 21667, ...?

||||

{ 470 }

Which number fits in the next spot in this sequence?:

228, 685, 2056, 6169, 18508, ...?

||||

{ 471 }

Which number fits in the next spot in this sequence?:

198, 397, 795, 1591, 3183, ...?

||||

{ 472 }

Which number fits in the next spot in this sequence?:

297, 595, 1191, 2383, 4767, ...?

||||

{ 473 }

Which number fits in the next spot in this sequence?:

4.1, 16.4, 65.6, 262.4, 1049.6, ...?

||||

{ 474 }

Which number fits in the next spot in this sequence?:

5.4, 16.2, 48.6, 145.8, 437.4, ...?

||||

{ 475 }

Solve this: what comes next in this sequence?:

109, 126, 143, 160, 177, ...?

||||

{ 476 }

Which number fits in the next spot in this sequence?:

2.3, 9.2, 36.8, 147.2, 588.8, ...?

||||

{ 477 }

Which number fits in the next spot in this sequence?:

296, 1185, 4741, 18965, 75861, ...?

||||

{ 478 }

Solve this: what comes next in this sequence?:

100, 123, 146, 169, 192, ...?

||||

{ 479 }

Which number fits in the next spot in this sequence?:

104, 313, 940, 2821, 8464, ...?

||||

{ 480 }

Which number fits in the next spot in this sequence?:

7, 28, 112, 448, 1792, ...?

||||

{ 481 }

Solve this: what comes next in this sequence?:

141, 189, 237, 285, 333, ...?

||||

{ 482 }

Which number fits in the next spot in this sequence?:

8.7, 17.4, 34.8, 69.6, 139.2, ...?

||||

{ 483 }

Which number fits in the next spot in this sequence?:

6, 24, 96, 384, 1536, ...?

||||

{ 484 }

Which number fits in the next spot in this sequence?:

226, 905, 3621, 14485, 57941, ...?

||||

{ 485 }

Which number fits in the next spot in this sequence?:

99, 199, 399, 799, 1599, ...?

||||

{ 486 }

Which number fits in the next spot in this sequence?:

153, 613, 2453, 9813, 39253, ...?

||||

{ 487 }

Solve this: what comes next in this sequence?:

236, 273, 310, 347, 384, ...?

||||

{ 488 }

Solve this: what comes next in this sequence?:

165, 213, 261, 309, 357, ...?

||||

{ 489 }

Which number fits in the next spot in this sequence?:

285, 856, 2569, 7708, 23125, ...?

||||

{ 490 }

Which number fits in the next spot in this sequence?:

291, 874, 2623, 7870, 23611, ...?

||||

{ 491 }

Solve this: what comes next in this sequence?:

229, 278, 327, 376, 425, ...?

||||

{ 492 }

Which number fits in the next spot in this sequence?:

9.7, 29.1, 87.3, 261.9, 785.7, ...?

||||

{ 493 }

Which number fits in the next spot in this sequence?:

7.5, 30, 120, 480, 1920, ...?

||||

{ 494 }

Which number fits in the next spot in this sequence?:

196, 393, 787, 1575, 3151, ...?

||||

{ 495 }

Solve this: what comes next in this sequence?:

103, 137, 171, 205, 239, ...?

||||

{ 496 }

Which number fits in the next spot in this sequence?:

8.7, 17.4, 34.8, 69.6, 139.2, ...?

||||

{ 497 }

Which number fits in the next spot in this sequence?:

151, 303, 607, 1215, 2431, ...?

||||

{ 498 }

Solve this: what comes next in this sequence?:

98, 127, 156, 185, 214, ...?

||||

{ 499 }

Solve this: what comes next in this sequence?:

145, 167, 189, 211, 233, ...?

||||

{ 500 }

Which number fits in the next spot in this sequence?:

100, 201, 403, 807, 1615, ...?

||||

{ 501 }

Which number fits in the next spot in this sequence?:

115, 346, 1039, 3118, 9355, ...?

||||

{ 502 }

Solve this: what comes next in this sequence?:

280, 299, 318, 337, 356, ...?

||||

{ 503 }

Which number fits in the next spot in this sequence?:

9, 18, 36, 72, 144, ...?

||||

{ 504 }

Which number fits in the next spot in this sequence?:

9.8, 29.4, 88.2, 264.6, 793.8, ...?

||||

{ 505 }

Solve this: what comes next in this sequence?:

197, 207, 217, 227, 237, ...?

||||

{ 506 }

Which number fits in the next spot in this sequence?:

180, 361, 723, 1447, 2895, ...?

||||

{ 507 }

Which number fits in the next spot in this sequence?:

141, 565, 2261, 9045, 36181, ...?

||||

{ 508 }

Solve this: what comes next in this sequence?:

127, 156, 185, 214, 243, ...?

||||

{ 509 }

Solve this: what comes next in this sequence?:

163, 193, 223, 253, 283, ...?

||||

{ 510 }

Which number fits in the next spot in this sequence?:

104, 417, 1669, 6677, 26709, ...?

||||

{ 511 }

Solve this: what comes next in this sequence?:

223, 257, 291, 325, 359, ...?

||||

{ 512 }

Which number fits in the next spot in this sequence?:

297, 892, 2677, 8032, 24097, ...?

||||

{ 513 }

Solve this: what comes next in this sequence?:

190, 207, 224, 241, 258, ...?

||||

{ 514 }

Solve this: what comes next in this sequence?:

96, 124, 152, 180, 208, ...?

||||

{ 515 }

Which number fits in the next spot in this sequence?:

6, 18, 54, 162, 486, ...?

||||

{ 516 }

Which number fits in the next spot in this sequence?:

121, 485, 1941, 7765, 31061, ...?

||||

{ 517 }

Which number fits in the next spot in this sequence?:

264, 793, 2380, 7141, 21424, ...?

||||

{ 518 }

Which number fits in the next spot in this sequence?:

151, 605, 2421, 9685, 38741, ...?

||||

{ 519 }

Which number fits in the next spot in this sequence?:

189, 757, 3029, 12117, 48469, ...?

||||

{ 520 }

Which number fits in the next spot in this sequence?:

4.3, 12.9, 38.7, 116.1, 348.3, ...?

||||

{ 521 }

Solve this: what comes next in this sequence?:

250, 294, 338, 382, 426, ...?

||||

{ 522 }

Which number fits in the next spot in this sequence?:

1.7, 6.8, 27.2, 108.8, 435.2, ...?

||||

{ 523 }

Which number fits in the next spot in this sequence?:

9.3, 37.2, 148.8, 595.2, 2380.8, ...?

||||

{ 524 }

Which number fits in the next spot in this sequence?:

110, 221, 443, 887, 1775, ...?

||||

{ 525 }

Which number fits in the next spot in this sequence?:

140, 281, 563, 1127, 2255, ...?

||||

{ 526 }

Which number fits in the next spot in this sequence?:

121, 364, 1093, 3280, 9841, ...?

||||

{ 527 }

Solve this: what comes next in this sequence?:

102, 119, 136, 153, 170, ...?

||||

{ 528 }

Which number fits in the next spot in this sequence?:

7.2, 21.6, 64.8, 194.4, 583.2, ...?

||||

{ 529 }

Which number fits in the next spot in this sequence?:

7.4, 29.6, 118.4, 473.6, 1894.4, ...?

||||

{ 530 }

Which number fits in the next spot in this sequence?:

112, 337, 1012, 3037, 9112, ...?

||||

{ 531 }

Which number fits in the next spot in this sequence?:

3.9, 15.6, 62.4, 249.6, 998.4, ...?

||||

{ 532 }

Solve this: what comes next in this sequence?:

250, 268, 286, 304, 322, ...?

||||

{ 533 }

Solve this: what comes next in this sequence?:

262, 273, 284, 295, 306, ...?

||||

{ 534 }

Which number fits in the next spot in this sequence?:

6.3, 18.9, 56.7, 170.1, 510.3, ...?

||||

{ 535 }

Which number fits in the next spot in this sequence?:

3.2, 12.8, 51.2, 204.8, 819.2, ...?

||||

{ 536 }

Which number fits in the next spot in this sequence?:

129, 517, 2069, 8277, 33109, ...?

||||

{ 537 }

Solve this: what comes next in this sequence?:

232, 282, 332, 382, 432, ...?

||||

{ 538 }

Solve this: what comes next in this sequence?:

251, 281, 311, 341, 371, ...?

||||

{ 539 }

Solve this: what comes next in this sequence?:

272, 293, 314, 335, 356, ...?

||||

{ 540 }

Solve this: what comes next in this sequence?:

106, 154, 202, 250, 298, ...?

||||

{ 541 }

Which number fits in the next spot in this sequence?:

122, 367, 1102, 3307, 9922, ...?

||||

{ 542 }

Which number fits in the next spot in this sequence?:

9.9, 19.8, 39.6, 79.2, 158.4, ...?

||||

{ 543 }

Solve this: what comes next in this sequence?:

244, 261, 278, 295, 312, ...?

||||

{ 544 }

Which number fits in the next spot in this sequence?:

5.6, 22.4, 89.6, 358.4, 1433.6, ...?

||||

{ 545 }

Which number fits in the next spot in this sequence?:

8.5, 34, 136, 544, 2176, ...?

||||

{ 546 }

Solve this: what comes next in this sequence?:

282, 308, 334, 360, 386, ...?

||||

{ 547 }

Which number fits in the next spot in this sequence?:

9.3, 37.2, 148.8, 595.2, 2380.8, ...?

||||

{ 548 }

Which number fits in the next spot in this sequence?:

4.6, 13.8, 41.4, 124.2, 372.6, ...?

||||

{ 549 }

Solve this: what comes next in this sequence?:

290, 331, 372, 413, 454, ...?

||||

{ 550 }

Solve this: what comes next in this sequence?:

118, 158, 198, 238, 278, ...?

||||

{ 551 }

Which number fits in the next spot in this sequence?:

3.7, 7.4, 14.8, 29.6, 59.2, ...?

||||

{ 552 }

Which number fits in the next spot in this sequence?:

5.3, 21.2, 84.8, 339.2, 1356.8, ...?

||||

{ 553 }

Which number fits in the next spot in this sequence?:

239, 479, 959, 1919, 3839, ...?

||||

{ 554 }

Which number fits in the next spot in this sequence?:

1.7, 5.1, 15.3, 45.9, 137.7, ...?

||||

{ 555 }

Solve this: what comes next in this sequence?:

199, 213, 227, 241, 255, ...?

||||

{ 556 }

Which number fits in the next spot in this sequence?:

7.9, 31.6, 126.4, 505.6, 2022.4, ...?

||||

{ 557 }

Which number fits in the next spot in this sequence?:

144, 289, 579, 1159, 2319, ...?

||||

{ 558 }

Solve this: what comes next in this sequence?:

285, 296, 307, 318, 329, ...?

||||

{ 559 }

Which number fits in the next spot in this sequence?:

1.8, 3.6, 7.2, 14.4, 28.8, ...?

||||

{ 560 }

Which number fits in the next spot in this sequence?:

1.3, 5.2, 20.8, 83.2, 332.8, ...?

||||

{ 561 }

Which number fits in the next spot in this sequence?:

6.7, 26.8, 107.2, 428.8, 1715.2, ...?

||||

{ 562 }

Which number fits in the next spot in this sequence?:

3.9, 15.6, 62.4, 249.6, 998.4, ...?

||||

{ 563 }

Which number fits in the next spot in this sequence?:

164, 657, 2629, 10517, 42069, ...?

||||

{ 564 }

Solve this: what comes next in this sequence?:

288, 331, 374, 417, 460, ...?

||||

{ 565 }

Solve this: what comes next in this sequence?:

124, 160, 196, 232, 268, ...?

||||

{ 566 }

Solve this: what comes next in this sequence?:

147, 173, 199, 225, 251, ...?

||||

{ 567 }

Which number fits in the next spot in this sequence?:

2, 6, 18, 54, 162, ...?

||||

{ 568 }

Solve this: what comes next in this sequence?:

236, 272, 308, 344, 380, ...?

||||

{ 569 }

Which number fits in the next spot in this sequence?:

230, 461, 923, 1847, 3695, ...?

||||

{ 570 }

Solve this: what comes next in this sequence?:

102, 145, 188, 231, 274, ...?

||||

{ 571 }

Which number fits in the next spot in this sequence?:

91, 183, 367, 735, 1471, ...?

||||

{ 572 }

Which number fits in the next spot in this sequence?:

149, 448, 1345, 4036, 12109, ...?

||||

{ 573 }

Solve this: what comes next in this sequence?:

90, 134, 178, 222, 266, ...?

||||

{ 574 }

Which number fits in the next spot in this sequence?:

233, 467, 935, 1871, 3743, ...?

||||

{ 575 }

Which number fits in the next spot in this sequence?:

9.5, 28.5, 85.5, 256.5, 769.5, ...?

||||

{ 576 }

Solve this: what comes next in this sequence?:

90, 131, 172, 213, 254, ...?

||||

{ 577 }

Which number fits in the next spot in this sequence?:

235, 706, 2119, 6358, 19075, ...?

||||

{ 578 }

Solve this: what comes next in this sequence?:

163, 208, 253, 298, 343, ...?

||||

{ 579 }

Which number fits in the next spot in this sequence?:

1.6, 3.2, 6.4, 12.8, 25.6, ...?

||||

{ 580 }

Which number fits in the next spot in this sequence?:

196, 785, 3141, 12565, 50261, ...?

||||

{ 581 }

Solve this: what comes next in this sequence?:

191, 202, 213, 224, 235, ...?

||||

{ 582 }

Which number fits in the next spot in this sequence?:

3.2, 12.8, 51.2, 204.8, 819.2, ...?

||||

{ 583 }

Solve this: what comes next in this sequence?:

143, 178, 213, 248, 283, ...?

||||

{ 584 }

Solve this: what comes next in this sequence?:

109, 154, 199, 244, 289, ...?

||||

{ 585 }

Solve this: what comes next in this sequence?:

253, 274, 295, 316, 337, ...?

||||

{ 586 }

Solve this: what comes next in this sequence?:

213, 259, 305, 351, 397, ...?

||||

{ 587 }

Which number fits in the next spot in this sequence?:

4.5, 9, 18, 36, 72, ...?

||||

{ 588 }

Which number fits in the next spot in this sequence?:

9.6, 19.2, 38.4, 76.8, 153.6, ...?

||||

{ 589 }

Which number fits in the next spot in this sequence?:

4.4, 17.6, 70.4, 281.6, 1126.4, ...?

||||

{ 590 }

Which number fits in the next spot in this sequence?:

115, 346, 1039, 3118, 9355, ...?

||||

{ 591 }

Which number fits in the next spot in this sequence?:

240, 721, 2164, 6493, 19480, ...?

||||

{ 592 }

Which number fits in the next spot in this sequence?:

100, 201, 403, 807, 1615, ...?

||||

{ 593 }

Which number fits in the next spot in this sequence?:

266, 799, 2398, 7195, 21586, ...?

||||

{ 594 }

Solve this: what comes next in this sequence?:

271, 306, 341, 376, 411, ...?

||||

{ 595 }

Which number fits in the next spot in this sequence?:

233, 933, 3733, 14933, 59733, ...?

||||

{ 596 }

Which number fits in the next spot in this sequence?:

1.3, 5.2, 20.8, 83.2, 332.8, ...?

||||

{ 597 }

Which number fits in the next spot in this sequence?:

214, 429, 859, 1719, 3439, ...?

||||

{ 598 }

Which number fits in the next spot in this sequence?:

275, 1101, 4405, 17621, 70485, ...?

||||

{ 599 }

Which number fits in the next spot in this sequence?:

3.7, 7.4, 14.8, 29.6, 59.2, ...?

||||

{ 600 }

Which number fits in the next spot in this sequence?:

9.3, 37.2, 148.8, 595.2, 2380.8, ...?

||||

{ 601 }

Which number fits in the next spot in this sequence?:

198, 595, 1786, 5359, 16078, ...?

||||

{ 602 }

Which number fits in the next spot in this sequence?:

159, 319, 639, 1279, 2559, ...?

||||

{ 603 }

Solve this: what comes next in this sequence?:

111, 132, 153, 174, 195, ...?

||||

{ 604 }

Which number fits in the next spot in this sequence?:

259, 778, 2335, 7006, 21019, ...?

||||

{ 605 }

Solve this: what comes next in this sequence?:

283, 317, 351, 385, 419, ...?

||||

{ 606 }

Which number fits in the next spot in this sequence?:

1.2, 2.4, 4.8, 9.6, 19.2, ...?

||||

{ 607 }

Which number fits in the next spot in this sequence?:

112, 337, 1012, 3037, 9112, ...?

||||

{ 608 }

Which number fits in the next spot in this sequence?:

185, 556, 1669, 5008, 15025, ...?

||||

{ 609 }

Which number fits in the next spot in this sequence?:

187, 749, 2997, 11989, 47957, ...?

||||

{ 610 }

Which number fits in the next spot in this sequence?:

131, 394, 1183, 3550, 10651, ...?

||||

{ 611 }

Solve this: what comes next in this sequence?:

182, 208, 234, 260, 286, ...?

||||

{ 612 }

Which number fits in the next spot in this sequence?:

9.8, 29.4, 88.2, 264.6, 793.8, ...?

||||

{ 613 }

Which number fits in the next spot in this sequence?:

204, 817, 3269, 13077, 52309, ...?

||||

{ 614 }

Which number fits in the next spot in this sequence?:

7, 28, 112, 448, 1792, ...?

||||

{ 615 }

Solve this: what comes next in this sequence?:

157, 193, 229, 265, 301, ...?

||||

{ 616 }

Which number fits in the next spot in this sequence?:

9.2, 18.4, 36.8, 73.6, 147.2, ...?

||||

{ 617 }

Which number fits in the next spot in this sequence?:

229, 459, 919, 1839, 3679, ...?

||||

{ 618 }

Which number fits in the next spot in this sequence?:

4, 12, 36, 108, 324, ...?

||||

{ 619 }

Which number fits in the next spot in this sequence?:

4.1, 8.2, 16.4, 32.8, 65.6, ...?

||||

{ 620 }

Solve this: what comes next in this sequence?:

231, 256, 281, 306, 331, ...?

||||

{ 621 }

Which number fits in the next spot in this sequence?:

142, 569, 2277, 9109, 36437, ...?

||||

{ 622 }

Solve this: what comes next in this sequence?:

167, 186, 205, 224, 243, ...?

||||

{ 623 }

Solve this: what comes next in this sequence?:

229, 240, 251, 262, 273, ...?

||||

{ 624 }

Which number fits in the next spot in this sequence?:

5.8, 11.6, 23.2, 46.4, 92.8000000000001, ...?

||||

{ 625 }

Which number fits in the next spot in this sequence?:

4.6, 18.4, 73.6, 294.4, 1177.6, ...?

||||

{ 626 }

Which number fits in the next spot in this sequence?:

214, 643, 1930, 5791, 17374, ...?

||||

{ 627 }

Solve this: what comes next in this sequence?:

189, 221, 253, 285, 317, ...?

||||

{ 628 }

Which number fits in the next spot in this sequence?:

265, 1061, 4245, 16981, 67925, ...?

||||

{ 629 }

Which number fits in the next spot in this sequence?:

1.3, 2.6, 5.2, 10.4, 20.8, ...?

||||

{ 630 }

Which number fits in the next spot in this sequence?:

165, 661, 2645, 10581, 42325, ...?

||||

{ 631 }

Solve this: what comes next in this sequence?:

232, 265, 298, 331, 364, ...?

||||

{ 632 }

Which number fits in the next spot in this sequence?:

6.3, 25.2, 100.8, 403.2, 1612.8, ...?

||||

{ 633 }

Solve this: what comes next in this sequence?:

97, 119, 141, 163, 185, ...?

||||

{ 634 }

Which number fits in the next spot in this sequence?:

4, 8, 16, 32, 64, ...?

||||

{ 635 }

Solve this: what comes next in this sequence?:

207, 248, 289, 330, 371, ...?

||||

{ 636 }

Solve this: what comes next in this sequence?:

120, 168, 216, 264, 312, ...?

||||

{ 637 }

Solve this: what comes next in this sequence?:

138, 176, 214, 252, 290, ...?

||||

{ 638 }

Which number fits in the next spot in this sequence?:

6.9, 27.6, 110.4, 441.6, 1766.4, ...?

||||

{ 639 }

Which number fits in the next spot in this sequence?:

243, 730, 2191, 6574, 19723, ...?

||||

{ 640 }

Solve this: what comes next in this sequence?:

298, 336, 374, 412, 450, ...?

||||

{ 641 }

Which number fits in the next spot in this sequence?:

237, 712, 2137, 6412, 19237, ...?

||||

{ 642 }

Solve this: what comes next in this sequence?:

229, 246, 263, 280, 297, ...?

||||

{ 643 }

Which number fits in the next spot in this sequence?:

249, 499, 999, 1999, 3999, ...?

||||

{ 644 }

Solve this: what comes next in this sequence?:

240, 257, 274, 291, 308, ...?

||||

{ 645 }

Which number fits in the next spot in this sequence?:

246, 493, 987, 1975, 3951, ...?

||||

{ 646 }

Which number fits in the next spot in this sequence?:

1.3, 5.2, 20.8, 83.2, 332.8, ...?

||||

{ 647 }

Which number fits in the next spot in this sequence?:

155, 466, 1399, 4198, 12595, ...?

||||

{ 648 }

Solve this: what comes next in this sequence?:

285, 314, 343, 372, 401, ...?

||||

{ 649 }

Solve this: what comes next in this sequence?:

194, 240, 286, 332, 378, ...?

||||

{ 650 }

Which number fits in the next spot in this sequence?:

197, 592, 1777, 5332, 15997, ...?

||||

{ 651 }

Which number fits in the next spot in this sequence?:

8.6, 34.4, 137.6, 550.4, 2201.6, ...?

||||

{ 652 }

Which number fits in the next spot in this sequence?:

299, 599, 1199, 2399, 4799, ...?

||||

{ 653 }

Which number fits in the next spot in this sequence?:

226, 905, 3621, 14485, 57941, ...?

||||

{ 654 }

Which number fits in the next spot in this sequence?:

223, 447, 895, 1791, 3583, ...?

||||

{ 655 }

Which number fits in the next spot in this sequence?:

4.1, 8.2, 16.4, 32.8, 65.6, ...?

||||

{ 656 }

Solve this: what comes next in this sequence?:

300, 322, 344, 366, 388, ...?

||||

{ 657 }

Which number fits in the next spot in this sequence?:

289, 1157, 4629, 18517, 74069, ...?

||||

{ 658 }

Which number fits in the next spot in this sequence?:

5.5, 16.5, 49.5, 148.5, 445.5, ...?

||||

{ 659 }

Which number fits in the next spot in this sequence?:

98, 393, 1573, 6293, 25173, ...?

||||

{ 660 }

Which number fits in the next spot in this sequence?:

7.2, 21.6, 64.8, 194.4, 583.2, ...?

||||

{ 661 }

Which number fits in the next spot in this sequence?:

4.8, 19.2, 76.8, 307.2, 1228.8, ...?

||||

{ 662 }

Which number fits in the next spot in this sequence?:

7.9, 31.6, 126.4, 505.6, 2022.4, ...?

||||

{ 663 }

Which number fits in the next spot in this sequence?:

3.8, 15.2, 60.8, 243.2, 972.8, ...?

||||

{ 664 }

Solve this: what comes next in this sequence?:

263, 311, 359, 407, 455, ...?

||||

{ 665 }

Which number fits in the next spot in this sequence?:

6.9, 13.8, 27.6, 55.2, 110.4, ...?

||||

{ 666 }

Solve this: what comes next in this sequence?:

117, 163, 209, 255, 301, ...?

||||

{ 667 }

Which number fits in the next spot in this sequence?:

213, 427, 855, 1711, 3423, ...?

||||

{ 668 }

Which number fits in the next spot in this sequence?:

199, 598, 1795, 5386, 16159, ...?

||||

{ 669 }

Solve this: what comes next in this sequence?:

230, 241, 252, 263, 274, ...?

||||

{ 670 }

Solve this: what comes next in this sequence?:

120, 149, 178, 207, 236, ...?

||||

{ 671 }

Which number fits in the next spot in this sequence?:

206, 825, 3301, 13205, 52821, ...?

||||

{ 672 }

Which number fits in the next spot in this sequence?:

2.4, 7.2, 21.6, 64.8, 194.4, ...?

||||

{ 673 }

Solve this: what comes next in this sequence?:

193, 243, 293, 343, 393, ...?

||||

{ 674 }

Which number fits in the next spot in this sequence?:

9.5, 38, 152, 608, 2432, ...?

||||

{ 675 }

Solve this: what comes next in this sequence?:

157, 185, 213, 241, 269, ...?

||||

{ 676 }

Which number fits in the next spot in this sequence?:

3.6, 10.8, 32.4, 97.2, 291.6, ...?

||||

{ 677 }

Which number fits in the next spot in this sequence?:

8.7, 34.8, 139.2, 556.8, 2227.2, ...?

||||

{ 678 }

Solve this: what comes next in this sequence?:

227, 237, 247, 257, 267, ...?

||||

{ 679 }

Solve this: what comes next in this sequence?:

105, 124, 143, 162, 181, ...?

||||

{ 680 }

Which number fits in the next spot in this sequence?:

259, 519, 1039, 2079, 4159, ...?

||||

{ 681 }

Which number fits in the next spot in this sequence?:

8.1, 16.2, 32.4, 64.8, 129.6, ...?

||||

{ 682 }

Solve this: what comes next in this sequence?:

197, 215, 233, 251, 269, ...?

||||

{ 683 }

Which number fits in the next spot in this sequence?:

9.1, 27.3, 81.9, 245.7, 737.1, ...?

||||

{ 684 }

Which number fits in the next spot in this sequence?:

4.4, 13.2, 39.6, 118.8, 356.4, ...?

||||

{ 685 }

Which number fits in the next spot in this sequence?:

220, 441, 883, 1767, 3535, ...?

||||

{ 686 }

Which number fits in the next spot in this sequence?:

4.6, 18.4, 73.6, 294.4, 1177.6, ...?

||||

{ 687 }

Which number fits in the next spot in this sequence?:

212, 425, 851, 1703, 3407, ...?

||||

{ 688 }

Which number fits in the next spot in this sequence?:

115, 346, 1039, 3118, 9355, ...?

||||

{ 689 }

Solve this: what comes next in this sequence?:

131, 181, 231, 281, 331, ...?

||||

{ 690 }

Solve this: what comes next in this sequence?:

185, 211, 237, 263, 289, ...?

||||

{ 691 }

Solve this: what comes next in this sequence?:

167, 185, 203, 221, 239, ...?

||||

{ 692 }

Which number fits in the next spot in this sequence?:

250, 501, 1003, 2007, 4015, ...?

||||

{ 693 }

Which number fits in the next spot in this sequence?:

6.3, 12.6, 25.2, 50.4, 100.8, ...?

||||

{ 694 }

Solve this: what comes next in this sequence?:

201, 232, 263, 294, 325, ...?

||||

{ 695 }

Which number fits in the next spot in this sequence?:

234, 469, 939, 1879, 3759, ...?

||||

{ 696 }

Which number fits in the next spot in this sequence?:

7, 14, 28, 56, 112, ...?

||||

{ 697 }

Which number fits in the next spot in this sequence?:

5.4, 21.6, 86.4, 345.6, 1382.4, ...?

||||

{ 698 }

Which number fits in the next spot in this sequence?:

9.3, 18.6, 37.2, 74.4, 148.8, ...?

||||

{ 699 }

Solve this: what comes next in this sequence?:

156, 197, 238, 279, 320, ...?

||||

{ 700 }

Which number fits in the next spot in this sequence?:

149, 448, 1345, 4036, 12109, ...?

||||

{ 701 }

Which number fits in the next spot in this sequence?:

7.5, 15, 30, 60, 120, ...?

||||

{ 702 }

Solve this: what comes next in this sequence?:

147, 161, 175, 189, 203, ...?

||||

{ 703 }

Solve this: what comes next in this sequence?:

190, 224, 258, 292, 326, ...?

||||

{ 704 }

Which number fits in the next spot in this sequence?:

162, 487, 1462, 4387, 13162, ...?

||||

{ 705 }

Solve this: what comes next in this sequence?:

116, 153, 190, 227, 264, ...?

||||

{ 706 }

Which number fits in the next spot in this sequence?:

109, 219, 439, 879, 1759, ...?

||||

{ 707 }

Which number fits in the next spot in this sequence?:

7.7, 30.8, 123.2, 492.8, 1971.2, ...?

||||

{ 708 }

Solve this: what comes next in this sequence?:

181, 194, 207, 220, 233, ...?

||||

{ 709 }

Solve this: what comes next in this sequence?:

240, 280, 320, 360, 400, ...?

||||

{ 710 }

Which number fits in the next spot in this sequence?:

292, 877, 2632, 7897, 23692, ...?

||||

{ 711 }

Which number fits in the next spot in this sequence?:

280, 561, 1123, 2247, 4495, ...?

||||

{ 712 }

Solve this: what comes next in this sequence?:

159, 207, 255, 303, 351, ...?

||||

{ 713 }

Which number fits in the next spot in this sequence?:

211, 423, 847, 1695, 3391, ...?

||||

{ 714 }

Which number fits in the next spot in this sequence?:

261, 523, 1047, 2095, 4191, ...?

||||

{ 715 }

Which number fits in the next spot in this sequence?:

159, 637, 2549, 10197, 40789, ...?

||||

{ 716 }

Solve this: what comes next in this sequence?:

140, 157, 174, 191, 208, ...?

||||

{ 717 }

Solve this: what comes next in this sequence?:

90, 114, 138, 162, 186, ...?

||||

{ 718 }

Which number fits in the next spot in this sequence?:

2.3, 9.2, 36.8, 147.2, 588.8, ...?

||||

{ 719 }

Solve this: what comes next in this sequence?:

221, 245, 269, 293, 317, ...?

||||

{ 720 }

Which number fits in the next spot in this sequence?:

3.3, 6.6, 13.2, 26.4, 52.8, ...?

||||

{ 721 }

Solve this: what comes next in this sequence?:

242, 267, 292, 317, 342, ...?

||||

{ 722 }

Which number fits in the next spot in this sequence?:

2.9, 5.8, 11.6, 23.2, 46.4, ...?

||||

{ 723 }

Which number fits in the next spot in this sequence?:

280, 1121, 4485, 17941, 71765, ...?

||||

{ 724 }

Which number fits in the next spot in this sequence?:

260, 1041, 4165, 16661, 66645, ...?

||||

{ 725 }

Solve this: what comes next in this sequence?:

134, 163, 192, 221, 250, ...?

||||

{ 726 }

Which number fits in the next spot in this sequence?:

9.9, 19.8, 39.6, 79.2, 158.4, ...?

||||

{ 727 }

Solve this: what comes next in this sequence?:

160, 193, 226, 259, 292, ...?

||||

{ 728 }

Which number fits in the next spot in this sequence?:

6.7, 13.4, 26.8, 53.6, 107.2, ...?

||||

{ 729 }

Which number fits in the next spot in this sequence?:

7, 21, 63, 189, 567, ...?

||||

{ 730 }

Which number fits in the next spot in this sequence?:

1.2, 4.8, 19.2, 76.8, 307.2, ...?

||||

{ 731 }

Solve this: what comes next in this sequence?:

118, 141, 164, 187, 210, ...?

||||

{ 732 }

Which number fits in the next spot in this sequence?:

6.7, 20.1, 60.3, 180.9, 542.7, ...?

||||

{ 733 }

Solve this: what comes next in this sequence?:

143, 154, 165, 176, 187, ...?

||||

{ 734 }

Which number fits in the next spot in this sequence?:

224, 897, 3589, 14357, 57429, ...?

||||

{ 735 }

Which number fits in the next spot in this sequence?:

7.1, 28.4, 113.6, 454.4, 1817.6, ...?

||||

{ 736 }

Which number fits in the next spot in this sequence?:

6.3, 18.9, 56.7, 170.1, 510.3, ...?

||||

{ 737 }

Which number fits in the next spot in this sequence?:

293, 880, 2641, 7924, 23773, ...?

||||

{ 738 }

Which number fits in the next spot in this sequence?:

117, 469, 1877, 7509, 30037, ...?

||||

{ 739 }

Which number fits in the next spot in this sequence?:

8.4, 16.8, 33.6, 67.2, 134.4, ...?

||||

{ 740 }

Which number fits in the next spot in this sequence?:

266, 1065, 4261, 17045, 68181, ...?

||||

{ 741 }

Solve this: what comes next in this sequence?:

180, 196, 212, 228, 244, ...?

||||

{ 742 }

Which number fits in the next spot in this sequence?:

116, 465, 1861, 7445, 29781, ...?

||||

{ 743 }

Solve this: what comes next in this sequence?:

144, 192, 240, 288, 336, ...?

||||

{ 744 }

Which number fits in the next spot in this sequence?:

3.3, 6.6, 13.2, 26.4, 52.8, ...?

||||

{ 745 }

Solve this: what comes next in this sequence?:

297, 308, 319, 330, 341, ...?

||||

{ 746 }

Which number fits in the next spot in this sequence?:

132, 397, 1192, 3577, 10732, ...?

||||

{ 747 }

Which number fits in the next spot in this sequence?:

140, 421, 1264, 3793, 11380, ...?

||||

{ 748 }

Which number fits in the next spot in this sequence?:

8.3, 16.6, 33.2, 66.4, 132.8, ...?

||||

{ 749 }

Which number fits in the next spot in this sequence?:

122, 245, 491, 983, 1967, ...?

||||

{ 750 }

Which number fits in the next spot in this sequence?:

1.3, 3.9, 11.7, 35.1, 105.3, ...?

||||

{ 751 }

Solve this: what comes next in this sequence?:

174, 189, 204, 219, 234, ...?

||||

{ 752 }

Which number fits in the next spot in this sequence?:

3.8, 11.4, 34.2, 102.6, 307.8, ...?

||||

{ 753 }

Which number fits in the next spot in this sequence?:

9.7, 38.8, 155.2, 620.8, 2483.2, ...?

||||

{ 754 }

Which number fits in the next spot in this sequence?:

223, 893, 3573, 14293, 57173, ...?

||||

{ 755 }

Which number fits in the next spot in this sequence?:

213, 853, 3413, 13653, 54613, ...?

||||

{ 756 }

Solve this: what comes next in this sequence?:

286, 322, 358, 394, 430, ...?

||||

{ 757 }

Solve this: what comes next in this sequence?:

124, 171, 218, 265, 312, ...?

||||

{ 758 }

Which number fits in the next spot in this sequence?:

115, 346, 1039, 3118, 9355, ...?

||||

{ 759 }

Which number fits in the next spot in this sequence?:

220, 881, 3525, 14101, 56405, ...?

||||

{ 760 }

Solve this: what comes next in this sequence?:

234, 256, 278, 300, 322, ...?

||||

{ 761 }

Which number fits in the next spot in this sequence?:

285, 856, 2569, 7708, 23125, ...?

||||

{ 762 }

Which number fits in the next spot in this sequence?:

3.5, 14, 56, 224, 896, ...?

||||

{ 763 }

Which number fits in the next spot in this sequence?:

3.6, 14.4, 57.6, 230.4, 921.6, ...?

||||

{ 764 }

Which number fits in the next spot in this sequence?:

207, 622, 1867, 5602, 16807, ...?

||||

{ 765 }

Solve this: what comes next in this sequence?:

169, 190, 211, 232, 253, ...?

||||

{ 766 }

Which number fits in the next spot in this sequence?:

192, 385, 771, 1543, 3087, ...?

||||

{ 767 }

Solve this: what comes next in this sequence?:

291, 336, 381, 426, 471, ...?

||||

{ 768 }

Which number fits in the next spot in this sequence?:

191, 383, 767, 1535, 3071, ...?

||||

{ 769 }

Which number fits in the next spot in this sequence?:

6.3, 18.9, 56.7, 170.1, 510.3, ...?

||||

{ 770 }

Which number fits in the next spot in this sequence?:

4.7, 18.8, 75.2, 300.8, 1203.2, ...?

||||

{ 771 }

Which number fits in the next spot in this sequence?:

233, 933, 3733, 14933, 59733, ...?

||||

{ 772 }

Solve this: what comes next in this sequence?:

111, 144, 177, 210, 243, ...?

||||

{ 773 }

Which number fits in the next spot in this sequence?:

8.1, 16.2, 32.4, 64.8, 129.6, ...?

||||

{ 774 }

Solve this: what comes next in this sequence?:

294, 344, 394, 444, 494, ...?

||||

{ 775 }

Which number fits in the next spot in this sequence?:

234, 469, 939, 1879, 3759, ...?

IIII

{ 776 }

Solve this: what comes next in this sequence?:

154, 191, 228, 265, 302, ...?

IIII

{ 777 }

Which number fits in the next spot in this sequence?:

264, 1057, 4229, 16917, 67669, ...?

IIII

{ 778 }

Which number fits in the next spot in this sequence?:

124, 497, 1989, 7957, 31829, ...?

IIII

{ 779 }

Which number fits in the next spot in this sequence?:

238, 953, 3813, 15253, 61013, ...?

IIII

{ 780 }

Which number fits in the next spot in this sequence?:

2.1, 6.3, 18.9, 56.7, 170.1, ...?

||||

{ 781 }

Solve this: what comes next in this sequence?:

291, 323, 355, 387, 419, ...?

||||

{ 782 }

Which number fits in the next spot in this sequence?:

285, 856, 2569, 7708, 23125, ...?

||||

{ 783 }

Which number fits in the next spot in this sequence?:

94, 283, 850, 2551, 7654, ...?

||||

{ 784 }

Which number fits in the next spot in this sequence?:

8.8, 17.6, 35.2, 70.4, 140.8, ...?

||||

{ 785 }

Solve this: what comes next in this sequence?:

193, 218, 243, 268, 293, ...?

||||

{ 786 }

Solve this: what comes next in this sequence?:

138, 174, 210, 246, 282, ...?

||||

{ 787 }

Which number fits in the next spot in this sequence?:

2.2, 8.8, 35.2, 140.8, 563.2, ...?

||||

{ 788 }

Which number fits in the next spot in this sequence?:

260, 781, 2344, 7033, 21100, ...?

||||

{ 789 }

Solve this: what comes next in this sequence?:

100, 119, 138, 157, 176, ...?

||||

{ 790 }

Solve this: what comes next in this sequence?:

223, 237, 251, 265, 279, ...?

||||

{ 791 }

Which number fits in the next spot in this sequence?:

263, 790, 2371, 7114, 21343, ...?

||||

{ 792 }

Which number fits in the next spot in this sequence?:

4.8, 19.2, 76.8, 307.2, 1228.8, ...?

||||

{ 793 }

Which number fits in the next spot in this sequence?:

4.9, 19.6, 78.4, 313.6, 1254.4, ...?

||||

{ 794 }

Solve this: what comes next in this sequence?:

196, 215, 234, 253, 272, ...?

||||

{ 795 }

Solve this: what comes next in this sequence?:

232, 270, 308, 346, 384, ...?

||||

{ 796 }

Solve this: what comes next in this sequence?:

94, 144, 194, 244, 294, ...?

||||

{ 797 }

Solve this: what comes next in this sequence?:

204, 217, 230, 243, 256, ...?

||||

{ 798 }

Which number fits in the next spot in this sequence?:

8.4, 16.8, 33.6, 67.2, 134.4, ...?

||||

{ 799 }

Which number fits in the next spot in this sequence?:

175, 351, 703, 1407, 2815, ...?

||||

{ 800 }

Solve this: what comes next in this sequence?:

178, 228, 278, 328, 378, ...?

||||

{ 801 }

Which number fits in the next spot in this sequence?:

2.7, 10.8, 43.2, 172.8, 691.2, ...?

||||

{ 802 }

Solve this: what comes next in this sequence?:

139, 160, 181, 202, 223, ...?

||||

{ 803 }

Which number fits in the next spot in this sequence?:

3.3, 13.2, 52.8, 211.2, 844.8, ...?

||||

{ 804 }

Which number fits in the next spot in this sequence?:

1.3, 5.2, 20.8, 83.2, 332.8, ...?

||||

{ 805 }

Which number fits in the next spot in this sequence?:

9.2, 36.8, 147.2, 588.8, 2355.2, ...?

||||

{ 806 }

Which number fits in the next spot in this sequence?:

114, 343, 1030, 3091, 9274, ...?

||||

{ 807 }

Which number fits in the next spot in this sequence?:

1.1, 3.3, 9.9, 29.7, 89.1000000000001, ...?

||||

{ 808 }

Which number fits in the next spot in this sequence?:

196, 393, 787, 1575, 3151, ...?

||||

{ 809 }

Which number fits in the next spot in this sequence?:

221, 664, 1993, 5980, 17941, ...?

||||

{ 810 }

Which number fits in the next spot in this sequence?:

217, 652, 1957, 5872, 17617, ...?

||||

{ 811 }

Which number fits in the next spot in this sequence?:

253, 507, 1015, 2031, 4063, ...?

||||

{ 812 }

Solve this: what comes next in this sequence?:

181, 223, 265, 307, 349, ...?

||||

{ 813 }

Which number fits in the next spot in this sequence?:

92, 185, 371, 743, 1487, ...?

||||

{ 814 }

Solve this: what comes next in this sequence?:

204, 231, 258, 285, 312, ...?

||||

{ 815 }

Solve this: what comes next in this sequence?:

197, 214, 231, 248, 265, ...?

||||

{ 816 }

Which number fits in the next spot in this sequence?:

198, 595, 1786, 5359, 16078, ...?

||||

{ 817 }

Which number fits in the next spot in this sequence?:

5.7, 22.8, 91.2, 364.8, 1459.2, ...?

||||

{ 818 }

Solve this: what comes next in this sequence?:

267, 315, 363, 411, 459, ...?

||||

{ 819 }

Which number fits in the next spot in this sequence?:

5.8, 17.4, 52.2, 156.6, 469.8, ...?

||||

{ 820 }

Solve this: what comes next in this sequence?:

91, 115, 139, 163, 187, ...?

||||

{ 421 }

Solve this: what comes next in this sequence?:

259, 308, 357, 406, 455, ...?

||||

By taking the differences between adjacent terms, we can see that this is an Arithmetic Progression, with common difference = 49. Therefore, the sixth term can be calculated by adding the common difference to the fifth term. It is equal to 455 + 49 = 504.

|||||||||

{ 422 }

Solve this: what comes next in this sequence?:

270, 319, 368, 417, 466, ...?

||||

By taking the differences between adjacent terms, we can see that this is an Arithmetic Progression, with common difference = 49. Therefore, the sixth term can be calculated by adding the common difference to the fifth term. It is equal to 466 + 49 = 515.

|||||||||

{ 423 }

Which number fits in the next spot in this sequence?:

6.4, 19.2, 57.6, 172.8, 518.4, ...?

IIII

By taking the ratios of adjacent terms, we can see that this is a

Geometric Progression, with common ratio = 3. Therefore, the sixth term

can be calculated by multiplying the fifth term with the common ratio. It is

equal to 518.4 x 3 = 1555.2.

IIIIIIII

{ 424 }

Which number fits in the next spot in this sequence?:

299, 599, 1199, 2399, 4799, ...?

IIII

By observation and trial and error, we find that each term in this

sequence is obtained by multiplying the previous term by 2 and adding 1.

Therefore, the sixth term = (2 x 4799) + 1 = 9599.

IIIIIIII

{ 425 }

Which number fits in the next spot in this sequence?:

7.1, 21.3, 63.9, 191.7, 575.1, ...?

IIII

By taking the ratios of adjacent terms, we can see that this is a Geometric Progression, with common ratio = 3. Therefore, the sixth term can be calculated by multiplying the fifth term with the common ratio. It is equal to 575.1 x 3 = 1725.3.

||||||||

{ 426 }

Which number fits in the next spot in this sequence?:

3.5, 10.5, 31.5, 94.5, 283.5, ...?

||||

By taking the ratios of adjacent terms, we can see that this is a Geometric Progression, with common ratio = 3. Therefore, the sixth term can be calculated by multiplying the fifth term with the common ratio. It is equal to 283.5 x 3 = 850.5.

|||||||||

{ 427 }

Solve this: what comes next in this sequence?:

150, 194, 238, 282, 326, ...?

||||

By taking the differences between adjacent terms, we can see that this is an Arithmetic Progression, with common difference = 44. Therefore, the sixth term can be calculated by adding the common difference to the fifth

term. It is equal to 326 + 44 = 370.

||||||||

{ 428 }

Solve this: what comes next in this sequence?:

217, 252, 287, 322, 357, ...?

||||

By taking the differences between adjacent terms, we can see that this is

an Arithmetic Progression, with common difference = 35. Therefore, the

sixth term can be calculated by adding the common difference to the fifth

term. It is equal to 357 + 35 = 392.

|||||||||

{ 429 }

Which number fits in the next spot in this sequence?:

92, 185, 371, 743, 1487, ...?

||||

By observation and trial and error, we find that each term in this

sequence is obtained by multiplying the previous term by 2 and adding 1.

Therefore, the sixth term = (2 x 1487) + 1 = 2975.

|||||||||

{ 430 }

Which number fits in the next spot in this sequence?:

142, 427, 1282, 3847, 11542, ...?

||||

By observation and trial and error, we find that each term in this

sequence is obtained by multiplying the previous term by 3 and adding 1.

Therefore, the sixth term = (3 x 11542) + 1 = 34627.

|||||||||

{ 431 }

Which number fits in the next spot in this sequence?:

6, 12, 24, 48, 96, ...?

||||

By taking the ratios of adjacent terms, we can see that this is a

Geometric Progression, with common ratio = 2. Therefore, the sixth term

can be calculated by multiplying the fifth term with the common ratio. It is

equal to 96 x 2 = 192.

||||||||

{ 432 }

Solve this: what comes next in this sequence?:

217, 260, 303, 346, 389, ...?

||||

By taking the differences between adjacent terms, we can see that this is an Arithmetic Progression, with common difference = 43. Therefore, the sixth term can be calculated by adding the common difference to the fifth term. It is equal to 389 + 43 = 432.

||||||||

{ 433 }

Solve this: what comes next in this sequence?:

221, 237, 253, 269, 285, ...?

||||

By taking the differences between adjacent terms, we can see that this is an Arithmetic Progression, with common difference = 16. Therefore, the sixth term can be calculated by adding the common difference to the fifth term. It is equal to 285 + 16 = 301.

|||||||||

{ 434 }

Solve this: what comes next in this sequence?:

216, 261, 306, 351, 396, ...?

||||

By taking the differences between adjacent terms, we can see that this is an Arithmetic Progression, with common difference = 45. Therefore, the sixth term can be calculated by adding the common difference to the fifth

term. It is equal to 396 + 45 = 441.

||||||||

{ 435 }

Which number fits in the next spot in this sequence?:

142, 569, 2277, 9109, 36437, ...?

||||

By observation and trial and error, we find that each term in this

sequence is obtained by multiplying the previous term by 4 and adding 1.

Therefore, the sixth term = (4 x 36437) + 1 = 145749.

|||||||||

{ 436 }

Which number fits in the next spot in this sequence?:

93, 187, 375, 751, 1503, ...?

||||

By observation and trial and error, we find that each term in this

sequence is obtained by multiplying the previous term by 2 and adding 1.

Therefore, the sixth term = (2 x 1503) + 1 = 3007.

||||||||

{ 437 }

Solve this: what comes next in this sequence?:

276, 312, 348, 384, 420, ...?

||||

By taking the differences between adjacent terms, we can see that this is

an Arithmetic Progression, with common difference = 36. Therefore, the

sixth term can be calculated by adding the common difference to the fifth

term. It is equal to 420 + 36 = 456.

|||||||||

{ 438 }

Solve this: what comes next in this sequence?:

94, 122, 150, 178, 206, ...?

||||

By taking the differences between adjacent terms, we can see that this is

an Arithmetic Progression, with common difference = 28. Therefore, the

sixth term can be calculated by adding the common difference to the fifth

term. It is equal to 206 + 28 = 234.

|||||||||

{ 439 }

Which number fits in the next spot in this sequence?:

153, 613, 2453, 9813, 39253, ...?

||||

By observation and trial and error, we find that each term in this sequence is obtained by multiplying the previous term by 4 and adding 1. Therefore, the sixth term = (4 x 39253) + 1 = 157013.

|||||||||

{ 440 }

Which number fits in the next spot in this sequence?:

268, 537, 1075, 2151, 4303, ...?

||||

By observation and trial and error, we find that each term in this sequence is obtained by multiplying the previous term by 2 and adding 1. Therefore, the sixth term = (2 x 4303) + 1 = 8607.

|||||||||

{ 441 }

Which number fits in the next spot in this sequence?:

2.8, 8.4, 25.2, 75.6, 226.8, ...?

||||

By taking the ratios of adjacent terms, we can see that this is a Geometric Progression, with common ratio = 3. Therefore, the sixth term can be calculated by multiplying the fifth term with the common ratio. It is equal to 226.8 x 3 = 680.4.

||||||||

{ 442 }

Which number fits in the next spot in this sequence?:

7.7, 15.4, 30.8, 61.6, 123.2, ...?

||||

By taking the ratios of adjacent terms, we can see that this is a

Geometric Progression, with common ratio = 2. Therefore, the sixth term

can be calculated by multiplying the fifth term with the common ratio. It is

equal to 123.2 x 2 = 246.4.

|||||||||

{ 443 }

Solve this: what comes next in this sequence?:

169, 208, 247, 286, 325, ...?

||||

By taking the differences between adjacent terms, we can see that this is

an Arithmetic Progression, with common difference = 39. Therefore, the

sixth term can be calculated by adding the common difference to the fifth

term. It is equal to 325 + 39 = 364.

|||||||||

{ 444 }

Which number fits in the next spot in this sequence?:

164, 657, 2629, 10517, 42069, ...?

||||

By observation and trial and error, we find that each term in this

sequence is obtained by multiplying the previous term by 4 and adding 1.

Therefore, the sixth term = (4 x 42069) + 1 = 168277.

|||||||||

{ 445 }

Which number fits in the next spot in this sequence?:

3.5, 14, 56, 224, 896, ...?

||||

By taking the ratios of adjacent terms, we can see that this is a

Geometric Progression, with common ratio = 4. Therefore, the sixth term

can be calculated by multiplying the fifth term with the common ratio. It is

equal to 896 x 4 = 3584.

|||||||||

{ 446 }

Solve this: what comes next in this sequence?:

177, 197, 217, 237, 257, ...?

||||

By taking the differences between adjacent terms, we can see that this is an Arithmetic Progression, with common difference = 20. Therefore, the sixth term can be calculated by adding the common difference to the fifth term. It is equal to 257 + 20 = 277.

|||||||||

{ 447 }

Which number fits in the next spot in this sequence?:

214, 857, 3429, 13717, 54869, ...?

||||

By observation and trial and error, we find that each term in this sequence is obtained by multiplying the previous term by 4 and adding 1. Therefore, the sixth term = (4 x 54869) + 1 = 219477.

|||||||||

{ 448 }

Which number fits in the next spot in this sequence?:

141, 565, 2261, 9045, 36181, ...?

||||

By observation and trial and error, we find that each term in this sequence is obtained by multiplying the previous term by 4 and adding 1. Therefore, the sixth term = (4 x 36181) + 1 = 144725.

|||||||||

{ 449 }

Which number fits in the next spot in this sequence?:

274, 549, 1099, 2199, 4399, ...?

IIII

By observation and trial and error, we find that each term in this

sequence is obtained by multiplying the previous term by 2 and adding 1.

Therefore, the sixth term = (2 x 4399) + 1 = 8799.

IIIIIIII

{ 450 }

Which number fits in the next spot in this sequence?:

4.7, 9.4, 18.8, 37.6, 75.2, ...?

IIII

By taking the ratios of adjacent terms, we can see that this is a

Geometric Progression, with common ratio = 2. Therefore, the sixth term

can be calculated by multiplying the fifth term with the common ratio. It is

equal to 75.2 x 2 = 150.4.

IIIIIIII

{ 451 }

Solve this: what comes next in this sequence?:

231, 275, 319, 363, 407, ...?

IIII

By taking the differences between adjacent terms, we can see that this is an Arithmetic Progression, with common difference = 44. Therefore, the sixth term can be calculated by adding the common difference to the fifth term. It is equal to 407 + 44 = 451.

|||||||||

{ 452 }

Which number fits in the next spot in this sequence?:

4.9, 19.6, 78.4, 313.6, 1254.4, ...?

||||

By taking the ratios of adjacent terms, we can see that this is a Geometric Progression, with common ratio = 4. Therefore, the sixth term can be calculated by multiplying the fifth term with the common ratio. It is equal to 1254.4 x 4 = 5017.6.

|||||||||

{ 453 }

Which number fits in the next spot in this sequence?:

182, 547, 1642, 4927, 14782, ...?

||||

By observation and trial and error, we find that each term in this sequence is obtained by multiplying the previous term by 3 and adding 1. Therefore, the sixth term = (3 x 14782) + 1 = 44347.

|||||||

{ 454 }

Solve this: what comes next in this sequence?:

249, 290, 331, 372, 413, ...?

||||

By taking the differences between adjacent terms, we can see that this is an Arithmetic Progression, with common difference = 41. Therefore, the sixth term can be calculated by adding the common difference to the fifth term. It is equal to 413 + 41 = 454.

|||||||||

{ 455 }

Solve this: what comes next in this sequence?:

111, 147, 183, 219, 255, ...?

||||

By taking the differences between adjacent terms, we can see that this is an Arithmetic Progression, with common difference = 36. Therefore, the sixth term can be calculated by adding the common difference to the fifth term. It is equal to 255 + 36 = 291.

|||||||||

{ 456 }

Which number fits in the next spot in this sequence?:

6, 18, 54, 162, 486, ...?

||||

By taking the ratios of adjacent terms, we can see that this is a

Geometric Progression, with common ratio = 3. Therefore, the sixth term

can be calculated by multiplying the fifth term with the common ratio. It is

equal to 486 x 3 = 1458.

|||||||||

{ 457 }

Which number fits in the next spot in this sequence?:

5.2, 10.4, 20.8, 41.6, 83.2, ...?

||||

By taking the ratios of adjacent terms, we can see that this is a

Geometric Progression, with common ratio = 2. Therefore, the sixth term

can be calculated by multiplying the fifth term with the common ratio. It is

equal to 83.2 x 2 = 166.4.

|||||||||

{ 458 }

Solve this: what comes next in this sequence?:

291, 314, 337, 360, 383, ...?

||||

By taking the differences between adjacent terms, we can see that this is an Arithmetic Progression, with common difference = 23. Therefore, the sixth term can be calculated by adding the common difference to the fifth term. It is equal to 383 + 23 = 406.

|||||||||

{ 459 }

Which number fits in the next spot in this sequence?:

3, 9, 27, 81, 243, ...?

||||

By taking the ratios of adjacent terms, we can see that this is a Geometric Progression, with common ratio = 3. Therefore, the sixth term can be calculated by multiplying the fifth term with the common ratio. It is equal to 243 x 3 = 729.

|||||||||

{ 460 }

Which number fits in the next spot in this sequence?:

204, 409, 819, 1639, 3279, ...?

||||

By observation and trial and error, we find that each term in this sequence is obtained by multiplying the previous term by 2 and adding 1.

Therefore, the sixth term = (2 x 3279) + 1 = 6559.

|||||||||

{ 461 }

Which number fits in the next spot in this sequence?:

5, 20, 80, 320, 1280, ...?

||||

By taking the ratios of adjacent terms, we can see that this is a

Geometric Progression, with common ratio = 4. Therefore, the sixth term

can be calculated by multiplying the fifth term with the common ratio. It is

equal to 1280 x 4 = 5120.

|||||||||

{ 462 }

Which number fits in the next spot in this sequence?:

165, 661, 2645, 10581, 42325, ...?

||||

By observation and trial and error, we find that each term in this

sequence is obtained by multiplying the previous term by 4 and adding 1.

Therefore, the sixth term = (4 x 42325) + 1 = 169301.

|||||||||

{ 463 }

Which number fits in the next spot in this sequence?:

188, 565, 1696, 5089, 15268, ...?

||||

By observation and trial and error, we find that each term in this

sequence is obtained by multiplying the previous term by 3 and adding 1.

Therefore, the sixth term = (3 x 15268) + 1 = 45805.

|||||||||

{ 464 }

Which number fits in the next spot in this sequence?:

2.9, 8.7, 26.1, 78.3, 234.9, ...?

||||

By taking the ratios of adjacent terms, we can see that this is a

Geometric Progression, with common ratio = 3. Therefore, the sixth term

can be calculated by multiplying the fifth term with the common ratio. It is

equal to 234.9 x 3 = 704.7.

|||||||||

{ 465 }

Which number fits in the next spot in this sequence?:

284, 1137, 4549, 18197, 72789, ...?

||||

By observation and trial and error, we find that each term in this sequence is obtained by multiplying the previous term by 4 and adding 1. Therefore, the sixth term = (4 x 72789) + 1 = 291157.

|||||||||

{ 466 }

Which number fits in the next spot in this sequence?:

160, 481, 1444, 4333, 13000, ...?

||||

By observation and trial and error, we find that each term in this sequence is obtained by multiplying the previous term by 3 and adding 1. Therefore, the sixth term = (3 x 13000) + 1 = 39001.

|||||||||

{ 467 }

Which number fits in the next spot in this sequence?:

7.5, 22.5, 67.5, 202.5, 607.5, ...?

||||

By taking the ratios of adjacent terms, we can see that this is a Geometric Progression, with common ratio = 3. Therefore, the sixth term can be calculated by multiplying the fifth term with the common ratio. It is equal to 607.5 x 3 = 1822.5.

|||||||||

{ 468 }

Which number fits in the next spot in this sequence?:

266, 1065, 4261, 17045, 68181, ...?

||||

By observation and trial and error, we find that each term in this

sequence is obtained by multiplying the previous term by 4 and adding 1.

Therefore, the sixth term = (4 x 68181) + 1 = 272725.

|||||||||

{ 469 }

Which number fits in the next spot in this sequence?:

267, 802, 2407, 7222, 21667, ...?

||||

By observation and trial and error, we find that each term in this

sequence is obtained by multiplying the previous term by 3 and adding 1.

Therefore, the sixth term = (3 x 21667) + 1 = 65002.

|||||||||

{ 470 }

Which number fits in the next spot in this sequence?:

228, 685, 2056, 6169, 18508, ...?

||||

By observation and trial and error, we find that each term in this

sequence is obtained by multiplying the previous term by 3 and adding 1.

Therefore, the sixth term = (3 x 18508) + 1 = 55525.

|||||||||

{ 471 }

Which number fits in the next spot in this sequence?:

198, 397, 795, 1591, 3183, ...?

||||

By observation and trial and error, we find that each term in this

sequence is obtained by multiplying the previous term by 2 and adding 1.

Therefore, the sixth term = (2 x 3183) + 1 = 6367.

|||||||||

{ 472 }

Which number fits in the next spot in this sequence?:

297, 595, 1191, 2383, 4767, ...?

||||

By observation and trial and error, we find that each term in this

sequence is obtained by multiplying the previous term by 2 and adding 1.

Therefore, the sixth term = (2 x 4767) + 1 = 9535.

|||||||||

{ 473 }

Which number fits in the next spot in this sequence?:

4.1, 16.4, 65.6, 262.4, 1049.6, ...?

||||

By taking the ratios of adjacent terms, we can see that this is a

Geometric Progression, with common ratio = 4. Therefore, the sixth term

can be calculated by multiplying the fifth term with the common ratio. It is

equal to 1049.6 x 4 = 4198.4.

|||||||||

{ 474 }

Which number fits in the next spot in this sequence?:

5.4, 16.2, 48.6, 145.8, 437.4, ...?

||||

By taking the ratios of adjacent terms, we can see that this is a

Geometric Progression, with common ratio = 3. Therefore, the sixth term

can be calculated by multiplying the fifth term with the common ratio. It is

equal to 437.4 x 3 = 1312.2.

|||||||||

{ 475 }

Solve this: what comes next in this sequence?:

109, 126, 143, 160, 177, ...?

||||

By taking the differences between adjacent terms, we can see that this is an Arithmetic Progression, with common difference = 17. Therefore, the sixth term can be calculated by adding the common difference to the fifth term. It is equal to 177 + 17 = 194.

|||||||||

{ 476 }

Which number fits in the next spot in this sequence?:

2.3, 9.2, 36.8, 147.2, 588.8, ...?

||||

By taking the ratios of adjacent terms, we can see that this is a Geometric Progression, with common ratio = 4. Therefore, the sixth term can be calculated by multiplying the fifth term with the common ratio. It is equal to 588.8 x 4 = 2355.2.

|||||||||

{ 477 }

Which number fits in the next spot in this sequence?:

296, 1185, 4741, 18965, 75861, ...?

||||

By observation and trial and error, we find that each term in this sequence is obtained by multiplying the previous term by 4 and adding 1.

Therefore, the sixth term = (4 x 75861) + 1 = 303445.

|||||||||

{ 478 }

Solve this: what comes next in this sequence?:

100, 123, 146, 169, 192, ...?

||||

By taking the differences between adjacent terms, we can see that this is an Arithmetic Progression, with common difference = 23. Therefore, the sixth term can be calculated by adding the common difference to the fifth term. It is equal to 192 + 23 = 215.

|||||||||

{ 479 }

Which number fits in the next spot in this sequence?:

104, 313, 940, 2821, 8464, ...?

||||

By observation and trial and error, we find that each term in this sequence is obtained by multiplying the previous term by 3 and adding 1. Therefore, the sixth term = (3 x 8464) + 1 = 25393.

|||||||||

{ 480 }

Which number fits in the next spot in this sequence?:

7, 28, 112, 448, 1792, ...?

||||

By taking the ratios of adjacent terms, we can see that this is a

Geometric Progression, with common ratio = 4. Therefore, the sixth term

can be calculated by multiplying the fifth term with the common ratio. It is

equal to 1792 x 4 = 7168.

|||||||||

{ 481 }

Solve this: what comes next in this sequence?:

141, 189, 237, 285, 333, ...?

||||

By taking the differences between adjacent terms, we can see that this is

an Arithmetic Progression, with common difference = 48. Therefore, the

sixth term can be calculated by adding the common difference to the fifth

term. It is equal to 333 + 48 = 381.

|||||||||

{ 482 }

Which number fits in the next spot in this sequence?:

8.7, 17.4, 34.8, 69.6, 139.2, ...?

||||

By taking the ratios of adjacent terms, we can see that this is a

Geometric Progression, with common ratio = 2. Therefore, the sixth term

can be calculated by multiplying the fifth term with the common ratio. It is

equal to 139.2 x 2 = 278.4.

|||||||||

{ 483 }

Which number fits in the next spot in this sequence?:

6, 24, 96, 384, 1536, ...?

||||

By taking the ratios of adjacent terms, we can see that this is a

Geometric Progression, with common ratio = 4. Therefore, the sixth term

can be calculated by multiplying the fifth term with the common ratio. It is

equal to 1536 x 4 = 6144.

||||||||

{ 484 }

Which number fits in the next spot in this sequence?:

226, 905, 3621, 14485, 57941, ...?

||||

By observation and trial and error, we find that each term in this

sequence is obtained by multiplying the previous term by 4 and adding 1.

Therefore, the sixth term = (4 x 57941) + 1 = 231765.

||||||||

{ 485 }

Which number fits in the next spot in this sequence?:

99, 199, 399, 799, 1599, ...?

||||

By observation and trial and error, we find that each term in this

sequence is obtained by multiplying the previous term by 2 and adding 1.

Therefore, the sixth term = (2 x 1599) + 1 = 3199.

||||||||

{ 486 }

Which number fits in the next spot in this sequence?:

153, 613, 2453, 9813, 39253, ...?

||||

By observation and trial and error, we find that each term in this

sequence is obtained by multiplying the previous term by 4 and adding 1.

Therefore, the sixth term = (4 x 39253) + 1 = 157013.

||||||||

{ 487 }

Solve this: what comes next in this sequence?:

236, 273, 310, 347, 384, ...?

||||

By taking the differences between adjacent terms, we can see that this is an Arithmetic Progression, with common difference = 37. Therefore, the sixth term can be calculated by adding the common difference to the fifth term. It is equal to 384 + 37 = 421.

|||||||||

{ 488 }

Solve this: what comes next in this sequence?:

165, 213, 261, 309, 357, ...?

||||

By taking the differences between adjacent terms, we can see that this is an Arithmetic Progression, with common difference = 48. Therefore, the sixth term can be calculated by adding the common difference to the fifth term. It is equal to 357 + 48 = 405.

|||||||||

{ 489 }

Which number fits in the next spot in this sequence?:

285, 856, 2569, 7708, 23125, ...?

||||

By observation and trial and error, we find that each term in this sequence is obtained by multiplying the previous term by 3 and adding 1. Therefore, the sixth term = (3 x 23125) + 1 = 69376.

|||||||||

{ 490 }

Which number fits in the next spot in this sequence?:

291, 874, 2623, 7870, 23611, ...?

||||

By observation and trial and error, we find that each term in this sequence is obtained by multiplying the previous term by 3 and adding 1. Therefore, the sixth term = (3 x 23611) + 1 = 70834.

||||||||

{ 491 }

Solve this: what comes next in this sequence?:

229, 278, 327, 376, 425, ...?

||||

By taking the differences between adjacent terms, we can see that this is an Arithmetic Progression, with common difference = 49. Therefore, the sixth term can be calculated by adding the common difference to the fifth term. It is equal to 425 + 49 = 474.

|||||||||

{ 492 }

Which number fits in the next spot in this sequence?:

9.7, 29.1, 87.3, 261.9, 785.7, ...?

||||

By taking the ratios of adjacent terms, we can see that this is a

Geometric Progression, with common ratio = 3. Therefore, the sixth term

can be calculated by multiplying the fifth term with the common ratio. It is

equal to 785.7 x 3 = 2357.1.

|||||||||

{ 493 }

Which number fits in the next spot in this sequence?:

7.5, 30, 120, 480, 1920, ...?

||||

By taking the ratios of adjacent terms, we can see that this is a

Geometric Progression, with common ratio = 4. Therefore, the sixth term

can be calculated by multiplying the fifth term with the common ratio. It is

equal to 1920 x 4 = 7680.

|||||||||

{ 494 }

Which number fits in the next spot in this sequence?:

196, 393, 787, 1575, 3151, ...?

||||

By observation and trial and error, we find that each term in this

sequence is obtained by multiplying the previous term by 2 and adding 1.

Therefore, the sixth term = (2 x 3151) + 1 = 6303.

|||||||||

{ 495 }

Solve this: what comes next in this sequence?:

103, 137, 171, 205, 239, ...?

||||

By taking the differences between adjacent terms, we can see that this is

an Arithmetic Progression, with common difference = 34. Therefore, the

sixth term can be calculated by adding the common difference to the fifth

term. It is equal to 239 + 34 = 273.

||||||||

{ 496 }

Which number fits in the next spot in this sequence?:

8.7, 17.4, 34.8, 69.6, 139.2, ...?

||||

By taking the ratios of adjacent terms, we can see that this is a Geometric Progression, with common ratio = 2. Therefore, the sixth term can be calculated by multiplying the fifth term with the common ratio. It is equal to 139.2 x 2 = 278.4.

|||||||||

{ 497 }

Which number fits in the next spot in this sequence?:

151, 303, 607, 1215, 2431, ...?

||||

By observation and trial and error, we find that each term in this sequence is obtained by multiplying the previous term by 2 and adding 1. Therefore, the sixth term = (2 x 2431) + 1 = 4863.

|||||||||

{ 498 }

Solve this: what comes next in this sequence?:

98, 127, 156, 185, 214, ...?

||||

By taking the differences between adjacent terms, we can see that this is an Arithmetic Progression, with common difference = 29. Therefore, the sixth term can be calculated by adding the common difference to the fifth term. It is equal to 214 + 29 = 243.

||||||||

{ 499 }

Solve this: what comes next in this sequence?:

145, 167, 189, 211, 233, ...?

||||

By taking the differences between adjacent terms, we can see that this is

an Arithmetic Progression, with common difference = 22. Therefore, the

sixth term can be calculated by adding the common difference to the fifth

term. It is equal to 233 + 22 = 255.

||||||||

{ 500 }

Which number fits in the next spot in this sequence?:

100, 201, 403, 807, 1615, ...?

||||

By observation and trial and error, we find that each term in this

sequence is obtained by multiplying the previous term by 2 and adding 1.

Therefore, the sixth term = (2 x 1615) + 1 = 3231.

||||||||

{ 501 }

Which number fits in the next spot in this sequence?:

115, 346, 1039, 3118, 9355, ...?

||||

By observation and trial and error, we find that each term in this

sequence is obtained by multiplying the previous term by 3 and adding 1.

Therefore, the sixth term = (3 x 9355) + 1 = 28066.

|||||||||

{ 502 }

Solve this: what comes next in this sequence?:

280, 299, 318, 337, 356, ...?

||||

By taking the differences between adjacent terms, we can see that this is

an Arithmetic Progression, with common difference = 19. Therefore, the

sixth term can be calculated by adding the common difference to the fifth

term. It is equal to 356 + 19 = 375.

|||||||||

{ 503 }

Which number fits in the next spot in this sequence?:

9, 18, 36, 72, 144, ...?

||||

By taking the ratios of adjacent terms, we can see that this is a

Geometric Progression, with common ratio = 2. Therefore, the sixth term

can be calculated by multiplying the fifth term with the common ratio. It is

equal to 144 x 2 = 288.

||||||||

{ 504 }

Which number fits in the next spot in this sequence?:

9.8, 29.4, 88.2, 264.6, 793.8, ...?

||||

By taking the ratios of adjacent terms, we can see that this is a

Geometric Progression, with common ratio = 3. Therefore, the sixth term

can be calculated by multiplying the fifth term with the common ratio. It is

equal to 793.8 x 3 = 2381.4.

||||||||

{ 505 }

Solve this: what comes next in this sequence?:

197, 207, 217, 227, 237, ...?

||||

By taking the differences between adjacent terms, we can see that this is

an Arithmetic Progression, with common difference = 10. Therefore, the

sixth term can be calculated by adding the common difference to the fifth

term. It is equal to 237 + 10 = 247.

||||||||

{ 506 }

Which number fits in the next spot in this sequence?:

180, 361, 723, 1447, 2895, ...?

||||

By observation and trial and error, we find that each term in this

sequence is obtained by multiplying the previous term by 2 and adding 1.

Therefore, the sixth term = (2 x 2895) + 1 = 5791.

||||||||

{ 507 }

Which number fits in the next spot in this sequence?:

141, 565, 2261, 9045, 36181, ...?

||||

By observation and trial and error, we find that each term in this

sequence is obtained by multiplying the previous term by 4 and adding 1.

Therefore, the sixth term = (4 x 36181) + 1 = 144725.

||||||||

{ 508 }

Solve this: what comes next in this sequence?:

127, 156, 185, 214, 243, ...?

||||

By taking the differences between adjacent terms, we can see that this is

an Arithmetic Progression, with common difference = 29. Therefore, the

sixth term can be calculated by adding the common difference to the fifth

term. It is equal to 243 + 29 = 272.

|||||||||

{ 509 }

Solve this: what comes next in this sequence?:

163, 193, 223, 253, 283, ...?

||||

By taking the differences between adjacent terms, we can see that this is

an Arithmetic Progression, with common difference = 30. Therefore, the

sixth term can be calculated by adding the common difference to the fifth

term. It is equal to 283 + 30 = 313.

|||||||||

{ 510 }

Which number fits in the next spot in this sequence?:

104, 417, 1669, 6677, 26709, ...?

||||

By observation and trial and error, we find that each term in this

sequence is obtained by multiplying the previous term by 4 and adding 1.

Therefore, the sixth term = (4 x 26709) + 1 = 106837.

||||||||

{ 511 }

Solve this: what comes next in this sequence?:

223, 257, 291, 325, 359, ...?

||||

By taking the differences between adjacent terms, we can see that this is

an Arithmetic Progression, with common difference = 34. Therefore, the

sixth term can be calculated by adding the common difference to the fifth

term. It is equal to 359 + 34 = 393.

|||||||||

{ 512 }

Which number fits in the next spot in this sequence?:

297, 892, 2677, 8032, 24097, ...?

||||

By observation and trial and error, we find that each term in this

sequence is obtained by multiplying the previous term by 3 and adding 1.

Therefore, the sixth term = (3 x 24097) + 1 = 72292.

|||||||

{ 513 }

Solve this: what comes next in this sequence?:

190, 207, 224, 241, 258, ...?

||||

By taking the differences between adjacent terms, we can see that this is

an Arithmetic Progression, with common difference = 17. Therefore, the

sixth term can be calculated by adding the common difference to the fifth

term. It is equal to 258 + 17 = 275.

|||||||

{ 514 }

Solve this: what comes next in this sequence?:

96, 124, 152, 180, 208, ...?

||||

By taking the differences between adjacent terms, we can see that this is

an Arithmetic Progression, with common difference = 28. Therefore, the

sixth term can be calculated by adding the common difference to the fifth

term. It is equal to 208 + 28 = 236.

|||||||

{ 515 }

Which number fits in the next spot in this sequence?:

6, 18, 54, 162, 486, ...?

||||

By taking the ratios of adjacent terms, we can see that this is a

Geometric Progression, with common ratio = 3. Therefore, the sixth term

can be calculated by multiplying the fifth term with the common ratio. It is

equal to 486 x 3 = 1458.

|||||||||

{ 516 }

Which number fits in the next spot in this sequence?:

121, 485, 1941, 7765, 31061, ...?

||||

By observation and trial and error, we find that each term in this

sequence is obtained by multiplying the previous term by 4 and adding 1.

Therefore, the sixth term = (4 x 31061) + 1 = 124245.

|||||||||

{ 517 }

Which number fits in the next spot in this sequence?:

264, 793, 2380, 7141, 21424, ...?

||||

By observation and trial and error, we find that each term in this sequence is obtained by multiplying the previous term by 3 and adding 1. Therefore, the sixth term = (3 x 21424) + 1 = 64273.

||||||||

{ 518 }

Which number fits in the next spot in this sequence?:

151, 605, 2421, 9685, 38741, ...?

||||

By observation and trial and error, we find that each term in this sequence is obtained by multiplying the previous term by 4 and adding 1. Therefore, the sixth term = (4 x 38741) + 1 = 154965.

||||||||

{ 519 }

Which number fits in the next spot in this sequence?:

189, 757, 3029, 12117, 48469, ...?

||||

By observation and trial and error, we find that each term in this sequence is obtained by multiplying the previous term by 4 and adding 1. Therefore, the sixth term = (4 x 48469) + 1 = 193877.

||||||||

{ 520 }

Which number fits in the next spot in this sequence?:

4.3, 12.9, 38.7, 116.1, 348.3, ...?

||||

By taking the ratios of adjacent terms, we can see that this is a

Geometric Progression, with common ratio = 3. Therefore, the sixth term

can be calculated by multiplying the fifth term with the common ratio. It is

equal to 348.3 x 3 = 1044.9.

|||||||||

{ 521 }

Solve this: what comes next in this sequence?:

250, 294, 338, 382, 426, ...?

||||

By taking the differences between adjacent terms, we can see that this is

an Arithmetic Progression, with common difference = 44. Therefore, the

sixth term can be calculated by adding the common difference to the fifth

term. It is equal to 426 + 44 = 470.

|||||||||

{ 522 }

Which number fits in the next spot in this sequence?:

1.7, 6.8, 27.2, 108.8, 435.2, ...?

||||

By taking the ratios of adjacent terms, we can see that this is a Geometric Progression, with common ratio = 4. Therefore, the sixth term can be calculated by multiplying the fifth term with the common ratio. It is equal to 435.2 x 4 = 1740.8.

|||||||||

{ 523 }

Which number fits in the next spot in this sequence?:

9.3, 37.2, 148.8, 595.2, 2380.8, ...?

||||

By taking the ratios of adjacent terms, we can see that this is a Geometric Progression, with common ratio = 4. Therefore, the sixth term can be calculated by multiplying the fifth term with the common ratio. It is equal to 2380.8 x 4 = 9523.2.

|||||||||

{ 524 }

Which number fits in the next spot in this sequence?:

110, 221, 443, 887, 1775, ...?

||||

By observation and trial and error, we find that each term in this sequence is obtained by multiplying the previous term by 2 and adding 1.

Therefore, the sixth term = (2 x 1775) + 1 = 3551.

IIIIIIII

{ 525 }

Which number fits in the next spot in this sequence?:

140, 281, 563, 1127, 2255, ...?

IIII

By observation and trial and error, we find that each term in this

sequence is obtained by multiplying the previous term by 2 and adding 1.

Therefore, the sixth term = (2 x 2255) + 1 = 4511.

IIIIIIII

{ 526 }

Which number fits in the next spot in this sequence?:

121, 364, 1093, 3280, 9841, ...?

IIII

By observation and trial and error, we find that each term in this

sequence is obtained by multiplying the previous term by 3 and adding 1.

Therefore, the sixth term = (3 x 9841) + 1 = 29524.

IIIIIIII

{ 527 }

Solve this: what comes next in this sequence?:

102, 119, 136, 153, 170, ...?

||||

By taking the differences between adjacent terms, we can see that this is

an Arithmetic Progression, with common difference = 17. Therefore, the

sixth term can be calculated by adding the common difference to the fifth

term. It is equal to 170 + 17 = 187.

|||||||||

{ 528 }

Which number fits in the next spot in this sequence?:

7.2, 21.6, 64.8, 194.4, 583.2, ...?

||||

By taking the ratios of adjacent terms, we can see that this is a

Geometric Progression, with common ratio = 3. Therefore, the sixth term

can be calculated by multiplying the fifth term with the common ratio. It is

equal to 583.2 x 3 = 1749.6.

|||||||||

{ 529 }

Which number fits in the next spot in this sequence?:

7.4, 29.6, 118.4, 473.6, 1894.4, ...?

||||

By taking the ratios of adjacent terms, we can see that this is a

Geometric Progression, with common ratio = 4. Therefore, the sixth term

can be calculated by multiplying the fifth term with the common ratio. It is

equal to 1894.4 x 4 = 7577.6.

|||||||||

{ 530 }

Which number fits in the next spot in this sequence?:

112, 337, 1012, 3037, 9112, ...?

||||

By observation and trial and error, we find that each term in this

sequence is obtained by multiplying the previous term by 3 and adding 1.

Therefore, the sixth term = (3 x 9112) + 1 = 27337.

|||||||||

{ 531 }

Which number fits in the next spot in this sequence?:

3.9, 15.6, 62.4, 249.6, 998.4, ...?

||||

By taking the ratios of adjacent terms, we can see that this is a

Geometric Progression, with common ratio = 4. Therefore, the sixth term

can be calculated by multiplying the fifth term with the common ratio. It is

equal to 998.4 x 4 = 3993.6.

IIIIIIIII

{ 532 }

Solve this: what comes next in this sequence?:

250, 268, 286, 304, 322, ...?

IIII

By taking the differences between adjacent terms, we can see that this is

an Arithmetic Progression, with common difference = 18. Therefore, the

sixth term can be calculated by adding the common difference to the fifth

term. It is equal to 322 + 18 = 340.

IIIIIIII

{ 533 }

Solve this: what comes next in this sequence?:

262, 273, 284, 295, 306, ...?

IIII

By taking the differences between adjacent terms, we can see that this is

an Arithmetic Progression, with common difference = 11. Therefore, the

sixth term can be calculated by adding the common difference to the fifth

term. It is equal to 306 + 11 = 317.

IIIIIIII

{ 534 }

Which number fits in the next spot in this sequence?:

6.3, 18.9, 56.7, 170.1, 510.3, ...?

||||

By taking the ratios of adjacent terms, we can see that this is a

Geometric Progression, with common ratio = 3. Therefore, the sixth term

can be calculated by multiplying the fifth term with the common ratio. It is

equal to 510.3 x 3 = 1530.9.

|||||||||

{ 535 }

Which number fits in the next spot in this sequence?:

3.2, 12.8, 51.2, 204.8, 819.2, ...?

||||

By taking the ratios of adjacent terms, we can see that this is a

Geometric Progression, with common ratio = 4. Therefore, the sixth term

can be calculated by multiplying the fifth term with the common ratio. It is

equal to 819.2 x 4 = 3276.8.

|||||||||

{ 536 }

Which number fits in the next spot in this sequence?:

129, 517, 2069, 8277, 33109, ...?

||||

By observation and trial and error, we find that each term in this sequence is obtained by multiplying the previous term by 4 and adding 1. Therefore, the sixth term = (4 x 33109) + 1 = 132437.

|||||||||

{ 537 }

Solve this: what comes next in this sequence?:

232, 282, 332, 382, 432, ...?

||||

By taking the differences between adjacent terms, we can see that this is an Arithmetic Progression, with common difference = 50. Therefore, the sixth term can be calculated by adding the common difference to the fifth term. It is equal to 432 + 50 = 482.

|||||||||

{ 538 }

Solve this: what comes next in this sequence?:

251, 281, 311, 341, 371, ...?

||||

By taking the differences between adjacent terms, we can see that this is an Arithmetic Progression, with common difference = 30. Therefore, the sixth term can be calculated by adding the common difference to the fifth

term. It is equal to 371 + 30 = 401.

IIIIIIII

{ 539 }

Solve this: what comes next in this sequence?:

272, 293, 314, 335, 356, ...?

IIII

By taking the differences between adjacent terms, we can see that this is

an Arithmetic Progression, with common difference = 21. Therefore, the

sixth term can be calculated by adding the common difference to the fifth

term. It is equal to 356 + 21 = 377.

IIIIIIII

{ 540 }

Solve this: what comes next in this sequence?:

106, 154, 202, 250, 298, ...?

IIII

By taking the differences between adjacent terms, we can see that this is

an Arithmetic Progression, with common difference = 48. Therefore, the

sixth term can be calculated by adding the common difference to the fifth

term. It is equal to 298 + 48 = 346.

IIIIIIII

{ 541 }

Which number fits in the next spot in this sequence?:

122, 367, 1102, 3307, 9922, ...?

||||

By observation and trial and error, we find that each term in this

sequence is obtained by multiplying the previous term by 3 and adding 1.

Therefore, the sixth term = (3 x 9922) + 1 = 29767.

|||||||||

{ 542 }

Which number fits in the next spot in this sequence?:

9.9, 19.8, 39.6, 79.2, 158.4, ...?

||||

By taking the ratios of adjacent terms, we can see that this is a

Geometric Progression, with common ratio = 2. Therefore, the sixth term

can be calculated by multiplying the fifth term with the common ratio. It is

equal to 158.4 x 2 = 316.8.

|||||||||

{ 543 }

Solve this: what comes next in this sequence?:

244, 261, 278, 295, 312, ...?

||||

By taking the differences between adjacent terms, we can see that this is an Arithmetic Progression, with common difference = 17. Therefore, the sixth term can be calculated by adding the common difference to the fifth term. It is equal to 312 + 17 = 329.

|||||||||

{ 544 }

Which number fits in the next spot in this sequence?:

5.6, 22.4, 89.6, 358.4, 1433.6, ...?

||||

By taking the ratios of adjacent terms, we can see that this is a Geometric Progression, with common ratio = 4. Therefore, the sixth term can be calculated by multiplying the fifth term with the common ratio. It is equal to 1433.6 x 4 = 5734.4.

|||||||||

{ 545 }

Which number fits in the next spot in this sequence?:

8.5, 34, 136, 544, 2176, ...?

||||

By taking the ratios of adjacent terms, we can see that this is a Geometric Progression, with common ratio = 4. Therefore, the sixth term can be calculated by multiplying the fifth term with the common ratio. It is

equal to 2176 x 4 = 8704.

||||||||

{ 546 }

Solve this: what comes next in this sequence?:

282, 308, 334, 360, 386, ...?

||||

By taking the differences between adjacent terms, we can see that this is

an Arithmetic Progression, with common difference = 26. Therefore, the

sixth term can be calculated by adding the common difference to the fifth

term. It is equal to 386 + 26 = 412.

||||||||

{ 547 }

Which number fits in the next spot in this sequence?:

9.3, 37.2, 148.8, 595.2, 2380.8, ...?

||||

By taking the ratios of adjacent terms, we can see that this is a

Geometric Progression, with common ratio = 4. Therefore, the sixth term

can be calculated by multiplying the fifth term with the common ratio. It is

equal to 2380.8 x 4 = 9523.2.

||||||||

{ 548 }

Which number fits in the next spot in this sequence?:

4.6, 13.8, 41.4, 124.2, 372.6, ...?

||||

By taking the ratios of adjacent terms, we can see that this is a Geometric Progression, with common ratio = 3. Therefore, the sixth term can be calculated by multiplying the fifth term with the common ratio. It is equal to 372.6 x 3 = 1117.8.

|||||||||

{ 549 }

Solve this: what comes next in this sequence?:

290, 331, 372, 413, 454, ...?

||||

By taking the differences between adjacent terms, we can see that this is an Arithmetic Progression, with common difference = 41. Therefore, the sixth term can be calculated by adding the common difference to the fifth term. It is equal to 454 + 41 = 495.

|||||||||

{ 550 }

Solve this: what comes next in this sequence?:

118, 158, 198, 238, 278, ...?

||||

By taking the differences between adjacent terms, we can see that this is an Arithmetic Progression, with common difference = 40. Therefore, the sixth term can be calculated by adding the common difference to the fifth term. It is equal to 278 + 40 = 318.

|||||||||

{ 551 }

Which number fits in the next spot in this sequence?:

3.7, 7.4, 14.8, 29.6, 59.2, ...?

||||

By taking the ratios of adjacent terms, we can see that this is a Geometric Progression, with common ratio = 2. Therefore, the sixth term can be calculated by multiplying the fifth term with the common ratio. It is equal to 59.2 x 2 = 118.4.

||||||||

{ 552 }

Which number fits in the next spot in this sequence?:

5.3, 21.2, 84.8, 339.2, 1356.8, ...?

||||

By taking the ratios of adjacent terms, we can see that this is a Geometric Progression, with common ratio = 4. Therefore, the sixth term

can be calculated by multiplying the fifth term with the common ratio. It is equal to 1356.8 x 4 = 5427.2.

||||||||

{ 553 }

Which number fits in the next spot in this sequence?:

239, 479, 959, 1919, 3839, ...?

||||

By observation and trial and error, we find that each term in this sequence is obtained by multiplying the previous term by 2 and adding 1. Therefore, the sixth term = (2 x 3839) + 1 = 7679.

||||||||

{ 554 }

Which number fits in the next spot in this sequence?:

1.7, 5.1, 15.3, 45.9, 137.7, ...?

||||

By taking the ratios of adjacent terms, we can see that this is a Geometric Progression, with common ratio = 3. Therefore, the sixth term can be calculated by multiplying the fifth term with the common ratio. It is equal to 137.7 x 3 = 413.1.

||||||||

{ 555 }

Solve this: what comes next in this sequence?:

199, 213, 227, 241, 255, ...?

||||

By taking the differences between adjacent terms, we can see that this is

an Arithmetic Progression, with common difference = 14. Therefore, the

sixth term can be calculated by adding the common difference to the fifth

term. It is equal to 255 + 14 = 269.

|||||||||

{ 556 }

Which number fits in the next spot in this sequence?:

7.9, 31.6, 126.4, 505.6, 2022.4, ...?

||||

By taking the ratios of adjacent terms, we can see that this is a

Geometric Progression, with common ratio = 4. Therefore, the sixth term

can be calculated by multiplying the fifth term with the common ratio. It is

equal to 2022.4 x 4 = 8089.6.

|||||||||

{ 557 }

Which number fits in the next spot in this sequence?:

144, 289, 579, 1159, 2319, ...?

||||

By observation and trial and error, we find that each term in this sequence is obtained by multiplying the previous term by 2 and adding 1. Therefore, the sixth term = (2 x 2319) + 1 = 4639.

|||||||||

{ 558 }

Solve this: what comes next in this sequence?:

285, 296, 307, 318, 329, ...?

||||

By taking the differences between adjacent terms, we can see that this is an Arithmetic Progression, with common difference = 11. Therefore, the sixth term can be calculated by adding the common difference to the fifth term. It is equal to 329 + 11 = 340.

|||||||||

{ 559 }

Which number fits in the next spot in this sequence?:

1.8, 3.6, 7.2, 14.4, 28.8, ...?

||||

By taking the ratios of adjacent terms, we can see that this is a Geometric Progression, with common ratio = 2. Therefore, the sixth term can be calculated by multiplying the fifth term with the common ratio. It is

equal to 28.8 x 2 = 57.6.

||||||||

{ 560 }

Which number fits in the next spot in this sequence?:

1.3, 5.2, 20.8, 83.2, 332.8, ...?

||||

By taking the ratios of adjacent terms, we can see that this is a

Geometric Progression, with common ratio = 4. Therefore, the sixth term

can be calculated by multiplying the fifth term with the common ratio. It is

equal to 332.8 x 4 = 1331.2.

||||||||

{ 561 }

Which number fits in the next spot in this sequence?:

6.7, 26.8, 107.2, 428.8, 1715.2, ...?

||||

By taking the ratios of adjacent terms, we can see that this is a

Geometric Progression, with common ratio = 4. Therefore, the sixth term

can be calculated by multiplying the fifth term with the common ratio. It is

equal to 1715.2 x 4 = 6860.8.

||||||||

{ 562 }

Which number fits in the next spot in this sequence?:

3.9, 15.6, 62.4, 249.6, 998.4, ...?

||||

By taking the ratios of adjacent terms, we can see that this is a Geometric Progression, with common ratio = 4. Therefore, the sixth term can be calculated by multiplying the fifth term with the common ratio. It is equal to 998.4 x 4 = 3993.6.

|||||||||

{ 563 }

Which number fits in the next spot in this sequence?:

164, 657, 2629, 10517, 42069, ...?

||||

By observation and trial and error, we find that each term in this sequence is obtained by multiplying the previous term by 4 and adding 1. Therefore, the sixth term = (4 x 42069) + 1 = 168277.

|||||||||

{ 564 }

Solve this: what comes next in this sequence?:

288, 331, 374, 417, 460, ...?

||||

By taking the differences between adjacent terms, we can see that this is an Arithmetic Progression, with common difference = 43. Therefore, the sixth term can be calculated by adding the common difference to the fifth term. It is equal to 460 + 43 = 503.

|||||||||

{ 565 }

Solve this: what comes next in this sequence?:

124, 160, 196, 232, 268, ...?

||||

By taking the differences between adjacent terms, we can see that this is an Arithmetic Progression, with common difference = 36. Therefore, the sixth term can be calculated by adding the common difference to the fifth term. It is equal to 268 + 36 = 304.

|||||||||

{ 566 }

Solve this: what comes next in this sequence?:

147, 173, 199, 225, 251, ...?

||||

By taking the differences between adjacent terms, we can see that this is an Arithmetic Progression, with common difference = 26. Therefore, the sixth term can be calculated by adding the common difference to the fifth

term. It is equal to 251 + 26 = 277.

||||||||

{ 567 }

Which number fits in the next spot in this sequence?:

2, 6, 18, 54, 162, ...?

||||

By taking the ratios of adjacent terms, we can see that this is a

Geometric Progression, with common ratio = 3. Therefore, the sixth term

can be calculated by multiplying the fifth term with the common ratio. It is

equal to 162 x 3 = 486.

||||||||

{ 568 }

Solve this: what comes next in this sequence?:

236, 272, 308, 344, 380, ...?

||||

By taking the differences between adjacent terms, we can see that this is

an Arithmetic Progression, with common difference = 36. Therefore, the

sixth term can be calculated by adding the common difference to the fifth

term. It is equal to 380 + 36 = 416.

||||||||

{ 569 }

Which number fits in the next spot in this sequence?:

230, 461, 923, 1847, 3695, ...?

||||

By observation and trial and error, we find that each term in this

sequence is obtained by multiplying the previous term by 2 and adding 1.

Therefore, the sixth term = (2 x 3695) + 1 = 7391.

|||||||||

{ 570 }

Solve this: what comes next in this sequence?:

102, 145, 188, 231, 274, ...?

||||

By taking the differences between adjacent terms, we can see that this is

an Arithmetic Progression, with common difference = 43. Therefore, the

sixth term can be calculated by adding the common difference to the fifth

term. It is equal to 274 + 43 = 317.

|||||||||

{ 571 }

Which number fits in the next spot in this sequence?:

91, 183, 367, 735, 1471, ...?

||||

By observation and trial and error, we find that each term in this sequence is obtained by multiplying the previous term by 2 and adding 1. Therefore, the sixth term = (2 x 1471) + 1 = 2943.

||||||||

{ 572 }

Which number fits in the next spot in this sequence?:

149, 448, 1345, 4036, 12109, ...?

||||

By observation and trial and error, we find that each term in this sequence is obtained by multiplying the previous term by 3 and adding 1. Therefore, the sixth term = (3 x 12109) + 1 = 36328.

||||||||

{ 573 }

Solve this: what comes next in this sequence?:

90, 134, 178, 222, 266, ...?

||||

By taking the differences between adjacent terms, we can see that this is an Arithmetic Progression, with common difference = 44. Therefore, the sixth term can be calculated by adding the common difference to the fifth term. It is equal to 266 + 44 = 310.

||||||||

{ 574 }

Which number fits in the next spot in this sequence?:

233, 467, 935, 1871, 3743, ...?

||||

By observation and trial and error, we find that each term in this

sequence is obtained by multiplying the previous term by 2 and adding 1.

Therefore, the sixth term = (2 x 3743) + 1 = 7487.

|||||||||

{ 575 }

Which number fits in the next spot in this sequence?:

9.5, 28.5, 85.5, 256.5, 769.5, ...?

||||

By taking the ratios of adjacent terms, we can see that this is a

Geometric Progression, with common ratio = 3. Therefore, the sixth term

can be calculated by multiplying the fifth term with the common ratio. It is

equal to 769.5 x 3 = 2308.5.

|||||||||

{ 576 }

Solve this: what comes next in this sequence?:

90, 131, 172, 213, 254, ...?

||||

By taking the differences between adjacent terms, we can see that this is an Arithmetic Progression, with common difference = 41. Therefore, the sixth term can be calculated by adding the common difference to the fifth term. It is equal to 254 + 41 = 295.

|||||||||

{ 577 }

Which number fits in the next spot in this sequence?:

235, 706, 2119, 6358, 19075, ...?

||||

By observation and trial and error, we find that each term in this sequence is obtained by multiplying the previous term by 3 and adding 1. Therefore, the sixth term = (3 x 19075) + 1 = 57226.

|||||||||

{ 578 }

Solve this: what comes next in this sequence?:

163, 208, 253, 298, 343, ...?

||||

By taking the differences between adjacent terms, we can see that this is an Arithmetic Progression, with common difference = 45. Therefore, the sixth term can be calculated by adding the common difference to the fifth term. It is equal to 343 + 45 = 388.

lIIIIIIII

{ 579 }

Which number fits in the next spot in this sequence?:

1.6, 3.2, 6.4, 12.8, 25.6, ...?

lIII

By taking the ratios of adjacent terms, we can see that this is a

Geometric Progression, with common ratio = 2. Therefore, the sixth term

can be calculated by multiplying the fifth term with the common ratio. It is

equal to 25.6 x 2 = 51.2.

lIIIIIIII

{ 580 }

Which number fits in the next spot in this sequence?:

196, 785, 3141, 12565, 50261, ...?

lIII

By observation and trial and error, we find that each term in this

sequence is obtained by multiplying the previous term by 4 and adding 1.

Therefore, the sixth term = (4 x 50261) + 1 = 201045.

lIIIIIIII

{ 581 }

Solve this: what comes next in this sequence?:

191, 202, 213, 224, 235, ...?

IIII

By taking the differences between adjacent terms, we can see that this is

an Arithmetic Progression, with common difference = 11. Therefore, the

sixth term can be calculated by adding the common difference to the fifth

term. It is equal to 235 + 11 = 246.

IIIIIIII

{ 582 }

Which number fits in the next spot in this sequence?:

3.2, 12.8, 51.2, 204.8, 819.2, ...?

IIII

By taking the ratios of adjacent terms, we can see that this is a

Geometric Progression, with common ratio = 4. Therefore, the sixth term

can be calculated by multiplying the fifth term with the common ratio. It is

equal to 819.2 x 4 = 3276.8.

IIIIIIII

{ 583 }

Solve this: what comes next in this sequence?:

143, 178, 213, 248, 283, ...?

||||

By taking the differences between adjacent terms, we can see that this is an Arithmetic Progression, with common difference = 35. Therefore, the sixth term can be calculated by adding the common difference to the fifth term. It is equal to 283 + 35 = 318.

|||||||||

{ 584 }

Solve this: what comes next in this sequence?:

109, 154, 199, 244, 289, ...?

||||

By taking the differences between adjacent terms, we can see that this is an Arithmetic Progression, with common difference = 45. Therefore, the sixth term can be calculated by adding the common difference to the fifth term. It is equal to 289 + 45 = 334.

|||||||||

{ 585 }

Solve this: what comes next in this sequence?:

253, 274, 295, 316, 337, ...?

||||

By taking the differences between adjacent terms, we can see that this is an Arithmetic Progression, with common difference = 21. Therefore, the

sixth term can be calculated by adding the common difference to the fifth

term. It is equal to 337 + 21 = 358.

||||||||

{ 586 }

Solve this: what comes next in this sequence?:

213, 259, 305, 351, 397, ...?

||||

By taking the differences between adjacent terms, we can see that this is

an Arithmetic Progression, with common difference = 46. Therefore, the

sixth term can be calculated by adding the common difference to the fifth

term. It is equal to 397 + 46 = 443.

||||||||

{ 587 }

Which number fits in the next spot in this sequence?:

4.5, 9, 18, 36, 72, ...?

||||

By taking the ratios of adjacent terms, we can see that this is a

Geometric Progression, with common ratio = 2. Therefore, the sixth term

can be calculated by multiplying the fifth term with the common ratio. It is

equal to 72 x 2 = 144.

||||||||

{ 588 }

Which number fits in the next spot in this sequence?:

9.6, 19.2, 38.4, 76.8, 153.6, ...?

||||

By taking the ratios of adjacent terms, we can see that this is a

Geometric Progression, with common ratio = 2. Therefore, the sixth term

can be calculated by multiplying the fifth term with the common ratio. It is

equal to 153.6 x 2 = 307.2.

|||||||||

{ 589 }

Which number fits in the next spot in this sequence?:

4.4, 17.6, 70.4, 281.6, 1126.4, ...?

||||

By taking the ratios of adjacent terms, we can see that this is a

Geometric Progression, with common ratio = 4. Therefore, the sixth term

can be calculated by multiplying the fifth term with the common ratio. It is

equal to 1126.4 x 4 = 4505.6.

|||||||||

{ 590 }

Which number fits in the next spot in this sequence?:

115, 346, 1039, 3118, 9355, ...?

||||

By observation and trial and error, we find that each term in this

sequence is obtained by multiplying the previous term by 3 and adding 1.

Therefore, the sixth term = (3 x 9355) + 1 = 28066.

|||||||||

{ 591 }

Which number fits in the next spot in this sequence?:

240, 721, 2164, 6493, 19480, ...?

||||

By observation and trial and error, we find that each term in this

sequence is obtained by multiplying the previous term by 3 and adding 1.

Therefore, the sixth term = (3 x 19480) + 1 = 58441.

|||||||||

{ 592 }

Which number fits in the next spot in this sequence?:

100, 201, 403, 807, 1615, ...?

||||

By observation and trial and error, we find that each term in this

sequence is obtained by multiplying the previous term by 2 and adding 1.

Therefore, the sixth term = (2 x 1615) + 1 = 3231.

|||||||||

{ 593 }

Which number fits in the next spot in this sequence?:

266, 799, 2398, 7195, 21586, ...?

||||

By observation and trial and error, we find that each term in this

sequence is obtained by multiplying the previous term by 3 and adding 1.

Therefore, the sixth term = (3 x 21586) + 1 = 64759.

|||||||||

{ 594 }

Solve this: what comes next in this sequence?:

271, 306, 341, 376, 411, ...?

||||

By taking the differences between adjacent terms, we can see that this is

an Arithmetic Progression, with common difference = 35. Therefore, the

sixth term can be calculated by adding the common difference to the fifth

term. It is equal to 411 + 35 = 446.

|||||||||

{ 595 }

Which number fits in the next spot in this sequence?:

233, 933, 3733, 14933, 59733, ...?

||||

By observation and trial and error, we find that each term in this sequence is obtained by multiplying the previous term by 4 and adding 1. Therefore, the sixth term = (4 x 59733) + 1 = 238933.

||||||||

{ 596 }

Which number fits in the next spot in this sequence?:

1.3, 5.2, 20.8, 83.2, 332.8, ...?

||||

By taking the ratios of adjacent terms, we can see that this is a Geometric Progression, with common ratio = 4. Therefore, the sixth term can be calculated by multiplying the fifth term with the common ratio. It is equal to 332.8 x 4 = 1331.2.

||||||||

{ 597 }

Which number fits in the next spot in this sequence?:

214, 429, 859, 1719, 3439, ...?

||||

By observation and trial and error, we find that each term in this sequence is obtained by multiplying the previous term by 2 and adding 1. Therefore, the sixth term = (2 x 3439) + 1 = 6879.

||||||||

{ 598 }

Which number fits in the next spot in this sequence?:

275, 1101, 4405, 17621, 70485, ...?

||||

By observation and trial and error, we find that each term in this sequence is obtained by multiplying the previous term by 4 and adding 1. Therefore, the sixth term = (4 x 70485) + 1 = 281941.

|||||||||

{ 599 }

Which number fits in the next spot in this sequence?:

3.7, 7.4, 14.8, 29.6, 59.2, ...?

||||

By taking the ratios of adjacent terms, we can see that this is a Geometric Progression, with common ratio = 2. Therefore, the sixth term can be calculated by multiplying the fifth term with the common ratio. It is equal to 59.2 x 2 = 118.4.

|||||||||

{ 600 }

Which number fits in the next spot in this sequence?:

9.3, 37.2, 148.8, 595.2, 2380.8, ...?

||||

By taking the ratios of adjacent terms, we can see that this is a Geometric Progression, with common ratio = 4. Therefore, the sixth term can be calculated by multiplying the fifth term with the common ratio. It is equal to 2380.8 x 4 = 9523.2.

|||||||||

{ 601 }

Which number fits in the next spot in this sequence?:

198, 595, 1786, 5359, 16078, ...?

||||

By observation and trial and error, we find that each term in this sequence is obtained by multiplying the previous term by 3 and adding 1. Therefore, the sixth term = (3 x 16078) + 1 = 48235.

|||||||||

{ 602 }

Which number fits in the next spot in this sequence?:

159, 319, 639, 1279, 2559, ...?

||||

By observation and trial and error, we find that each term in this sequence is obtained by multiplying the previous term by 2 and adding 1. Therefore, the sixth term = (2 x 2559) + 1 = 5119.

|||||||||

{ 603 }

Solve this: what comes next in this sequence?:

111, 132, 153, 174, 195, ...?

IIII

By taking the differences between adjacent terms, we can see that this is

an Arithmetic Progression, with common difference = 21. Therefore, the

sixth term can be calculated by adding the common difference to the fifth

term. It is equal to 195 + 21 = 216.

IIIIIIII

{ 604 }

Which number fits in the next spot in this sequence?:

259, 778, 2335, 7006, 21019, ...?

IIII

By observation and trial and error, we find that each term in this

sequence is obtained by multiplying the previous term by 3 and adding 1.

Therefore, the sixth term = (3 x 21019) + 1 = 63058.

IIIIIIII

{ 605 }

Solve this: what comes next in this sequence?:

283, 317, 351, 385, 419, ...?

IIII

By taking the differences between adjacent terms, we can see that this is an Arithmetic Progression, with common difference = 34. Therefore, the sixth term can be calculated by adding the common difference to the fifth term. It is equal to 419 + 34 = 453.

||||||||

{ 606 }

Which number fits in the next spot in this sequence?:

1.2, 2.4, 4.8, 9.6, 19.2, ...?

||||

By taking the ratios of adjacent terms, we can see that this is a Geometric Progression, with common ratio = 2. Therefore, the sixth term can be calculated by multiplying the fifth term with the common ratio. It is equal to 19.2 x 2 = 38.4.

||||||||

{ 607 }

Which number fits in the next spot in this sequence?:

112, 337, 1012, 3037, 9112, ...?

||||

By observation and trial and error, we find that each term in this sequence is obtained by multiplying the previous term by 3 and adding 1. Therefore, the sixth term = (3 x 9112) + 1 = 27337.

IIIIIIIII

{ 608 }

Which number fits in the next spot in this sequence?:

185, 556, 1669, 5008, 15025, ...?

IIII

By observation and trial and error, we find that each term in this

sequence is obtained by multiplying the previous term by 3 and adding 1.

Therefore, the sixth term = (3 x 15025) + 1 = 45076.

IIIIIIIII

{ 609 }

Which number fits in the next spot in this sequence?:

187, 749, 2997, 11989, 47957, ...?

IIII

By observation and trial and error, we find that each term in this

sequence is obtained by multiplying the previous term by 4 and adding 1.

Therefore, the sixth term = (4 x 47957) + 1 = 191829.

IIIIIIIII

{ 610 }

Which number fits in the next spot in this sequence?:

131, 394, 1183, 3550, 10651, ...?

IIII

By observation and trial and error, we find that each term in this

sequence is obtained by multiplying the previous term by 3 and adding 1.

Therefore, the sixth term = (3 x 10651) + 1 = 31954.

IIIIIIII

{ 611 }

Solve this: what comes next in this sequence?:

182, 208, 234, 260, 286, ...?

IIII

By taking the differences between adjacent terms, we can see that this is

an Arithmetic Progression, with common difference = 26. Therefore, the

sixth term can be calculated by adding the common difference to the fifth

term. It is equal to 286 + 26 = 312.

IIIIIIII

{ 612 }

Which number fits in the next spot in this sequence?:

9.8, 29.4, 88.2, 264.6, 793.8, ...?

IIII

By taking the ratios of adjacent terms, we can see that this is a Geometric Progression, with common ratio = 3. Therefore, the sixth term can be calculated by multiplying the fifth term with the common ratio. It is equal to 793.8 x 3 = 2381.4.

||||||||

{ 613 }

Which number fits in the next spot in this sequence?:

204, 817, 3269, 13077, 52309, ...?

||||

By observation and trial and error, we find that each term in this sequence is obtained by multiplying the previous term by 4 and adding 1. Therefore, the sixth term = (4 x 52309) + 1 = 209237.

|||||||||

{ 614 }

Which number fits in the next spot in this sequence?:

7, 28, 112, 448, 1792, ...?

||||

By taking the ratios of adjacent terms, we can see that this is a Geometric Progression, with common ratio = 4. Therefore, the sixth term can be calculated by multiplying the fifth term with the common ratio. It is equal to 1792 x 4 = 7168.

||||||||

{ 615 }

Solve this: what comes next in this sequence?:

157, 193, 229, 265, 301, ...?

||||

By taking the differences between adjacent terms, we can see that this is

an Arithmetic Progression, with common difference = 36. Therefore, the

sixth term can be calculated by adding the common difference to the fifth

term. It is equal to 301 + 36 = 337.

||||||||

{ 616 }

Which number fits in the next spot in this sequence?:

9.2, 18.4, 36.8, 73.6, 147.2, ...?

||||

By taking the ratios of adjacent terms, we can see that this is a

Geometric Progression, with common ratio = 2. Therefore, the sixth term

can be calculated by multiplying the fifth term with the common ratio. It is

equal to 147.2 x 2 = 294.4.

||||||||

{ 617 }

Which number fits in the next spot in this sequence?:

229, 459, 919, 1839, 3679, ...?

||||

By observation and trial and error, we find that each term in this sequence is obtained by multiplying the previous term by 2 and adding 1. Therefore, the sixth term = (2 x 3679) + 1 = 7359.

|||||||||

{ 618 }

Which number fits in the next spot in this sequence?:

4, 12, 36, 108, 324, ...?

||||

By taking the ratios of adjacent terms, we can see that this is a Geometric Progression, with common ratio = 3. Therefore, the sixth term can be calculated by multiplying the fifth term with the common ratio. It is equal to 324 x 3 = 972.

|||||||||

{ 619 }

Which number fits in the next spot in this sequence?:

4.1, 8.2, 16.4, 32.8, 65.6, ...?

||||

By taking the ratios of adjacent terms, we can see that this is a Geometric Progression, with common ratio = 2. Therefore, the sixth term can be calculated by multiplying the fifth term with the common ratio. It is equal to 65.6 x 2 = 131.2.

|||||||||

{ 620 }

Solve this: what comes next in this sequence?:

231, 256, 281, 306, 331, ...?

||||

By taking the differences between adjacent terms, we can see that this is an Arithmetic Progression, with common difference = 25. Therefore, the sixth term can be calculated by adding the common difference to the fifth term. It is equal to 331 + 25 = 356.

|||||||||

{ 621 }

Which number fits in the next spot in this sequence?:

142, 569, 2277, 9109, 36437, ...?

||||

By observation and trial and error, we find that each term in this sequence is obtained by multiplying the previous term by 4 and adding 1. Therefore, the sixth term = (4 x 36437) + 1 = 145749.

|||||||||

{ 622 }

Solve this: what comes next in this sequence?:

167, 186, 205, 224, 243, ...?

||||

By taking the differences between adjacent terms, we can see that this is an Arithmetic Progression, with common difference = 19. Therefore, the sixth term can be calculated by adding the common difference to the fifth term. It is equal to 243 + 19 = 262.

|||||||||

{ 623 }

Solve this: what comes next in this sequence?:

229, 240, 251, 262, 273, ...?

||||

By taking the differences between adjacent terms, we can see that this is an Arithmetic Progression, with common difference = 11. Therefore, the sixth term can be calculated by adding the common difference to the fifth term. It is equal to 273 + 11 = 284.

|||||||||

{ 624 }

Which number fits in the next spot in this sequence?:

5.8, 11.6, 23.2, 46.4, 92.8000000000001, ...?

||||

By taking the ratios of adjacent terms, we can see that this is a

Geometric Progression, with common ratio = 2. Therefore, the sixth term

can be calculated by multiplying the fifth term with the common ratio. It is

equal to 92.8000000000001 x 2 = 185.6.

|||||||||

{ 625 }

Which number fits in the next spot in this sequence?:

4.6, 18.4, 73.6, 294.4, 1177.6, ...?

||||

By taking the ratios of adjacent terms, we can see that this is a

Geometric Progression, with common ratio = 4. Therefore, the sixth term

can be calculated by multiplying the fifth term with the common ratio. It is

equal to 1177.6 x 4 = 4710.4.

|||||||||

{ 626 }

Which number fits in the next spot in this sequence?:

214, 643, 1930, 5791, 17374, ...?

||||

By observation and trial and error, we find that each term in this sequence is obtained by multiplying the previous term by 3 and adding 1. Therefore, the sixth term = (3 x 17374) + 1 = 52123.

|||||||||

{ 627 }

Solve this: what comes next in this sequence?:

189, 221, 253, 285, 317, ...?

||||

By taking the differences between adjacent terms, we can see that this is an Arithmetic Progression, with common difference = 32. Therefore, the sixth term can be calculated by adding the common difference to the fifth term. It is equal to 317 + 32 = 349.

|||||||||

{ 628 }

Which number fits in the next spot in this sequence?:

265, 1061, 4245, 16981, 67925, ...?

||||

By observation and trial and error, we find that each term in this sequence is obtained by multiplying the previous term by 4 and adding 1. Therefore, the sixth term = (4 x 67925) + 1 = 271701.

IIIIIIII

{ 629 }

Which number fits in the next spot in this sequence?:

1.3, 2.6, 5.2, 10.4, 20.8, ...?

IIII

By taking the ratios of adjacent terms, we can see that this is a

Geometric Progression, with common ratio = 2. Therefore, the sixth term

can be calculated by multiplying the fifth term with the common ratio. It is

equal to 20.8 x 2 = 41.6.

IIIIIIII

{ 630 }

Which number fits in the next spot in this sequence?:

165, 661, 2645, 10581, 42325, ...?

IIII

By observation and trial and error, we find that each term in this

sequence is obtained by multiplying the previous term by 4 and adding 1.

Therefore, the sixth term = (4 x 42325) + 1 = 169301.

IIIIIIII

{ 631 }

Solve this: what comes next in this sequence?:

232, 265, 298, 331, 364, ...?

||||

By taking the differences between adjacent terms, we can see that this is

an Arithmetic Progression, with common difference = 33. Therefore, the

sixth term can be calculated by adding the common difference to the fifth

term. It is equal to 364 + 33 = 397.

|||||||||

{ 632 }

Which number fits in the next spot in this sequence?:

6.3, 25.2, 100.8, 403.2, 1612.8, ...?

||||

By taking the ratios of adjacent terms, we can see that this is a

Geometric Progression, with common ratio = 4. Therefore, the sixth term

can be calculated by multiplying the fifth term with the common ratio. It is

equal to 1612.8 x 4 = 6451.2.

|||||||||

{ 633 }

Solve this: what comes next in this sequence?:

97, 119, 141, 163, 185, ...?

||||

By taking the differences between adjacent terms, we can see that this is

an Arithmetic Progression, with common difference = 22. Therefore, the

sixth term can be calculated by adding the common difference to the fifth

term. It is equal to 185 + 22 = 207.

|||||||||

{ 634 }

Which number fits in the next spot in this sequence?:

4, 8, 16, 32, 64, ...?

||||

By taking the ratios of adjacent terms, we can see that this is a

Geometric Progression, with common ratio = 2. Therefore, the sixth term

can be calculated by multiplying the fifth term with the common ratio. It is

equal to 64 x 2 = 128.

|||||||||

{ 635 }

Solve this: what comes next in this sequence?:

207, 248, 289, 330, 371, ...?

||||

By taking the differences between adjacent terms, we can see that this is

an Arithmetic Progression, with common difference = 41. Therefore, the

sixth term can be calculated by adding the common difference to the fifth term. It is equal to 371 + 41 = 412.

||||||||||

{ 636 }

Solve this: what comes next in this sequence?:

120, 168, 216, 264, 312, ...?

||||

By taking the differences between adjacent terms, we can see that this is an Arithmetic Progression, with common difference = 48. Therefore, the sixth term can be calculated by adding the common difference to the fifth term. It is equal to 312 + 48 = 360.

||||||||||

{ 637 }

Solve this: what comes next in this sequence?:

138, 176, 214, 252, 290, ...?

||||

By taking the differences between adjacent terms, we can see that this is an Arithmetic Progression, with common difference = 38. Therefore, the sixth term can be calculated by adding the common difference to the fifth term. It is equal to 290 + 38 = 328.

||||||||||

{ 638 }

Which number fits in the next spot in this sequence?:

6.9, 27.6, 110.4, 441.6, 1766.4, ...?

||||

By taking the ratios of adjacent terms, we can see that this is a

Geometric Progression, with common ratio = 4. Therefore, the sixth term

can be calculated by multiplying the fifth term with the common ratio. It is

equal to 1766.4 x 4 = 7065.6.

|||||||||

{ 639 }

Which number fits in the next spot in this sequence?:

243, 730, 2191, 6574, 19723, ...?

||||

By observation and trial and error, we find that each term in this

sequence is obtained by multiplying the previous term by 3 and adding 1.

Therefore, the sixth term = (3 x 19723) + 1 = 59170.

|||||||||

{ 640 }

Solve this: what comes next in this sequence?:

298, 336, 374, 412, 450, ...?

||||

By taking the differences between adjacent terms, we can see that this is an Arithmetic Progression, with common difference = 38. Therefore, the sixth term can be calculated by adding the common difference to the fifth term. It is equal to 450 + 38 = 488.

IIIIIIII

{ 641 }

Which number fits in the next spot in this sequence?:

237, 712, 2137, 6412, 19237, ...?

IIII

By observation and trial and error, we find that each term in this sequence is obtained by multiplying the previous term by 3 and adding 1. Therefore, the sixth term = (3 x 19237) + 1 = 57712.

IIIIIIII

{ 642 }

Solve this: what comes next in this sequence?:

229, 246, 263, 280, 297, ...?

IIII

By taking the differences between adjacent terms, we can see that this is an Arithmetic Progression, with common difference = 17. Therefore, the sixth term can be calculated by adding the common difference to the fifth term. It is equal to 297 + 17 = 314.

|||||||||

{ 643 }

Which number fits in the next spot in this sequence?:

249, 499, 999, 1999, 3999, ...?

||||

By observation and trial and error, we find that each term in this

sequence is obtained by multiplying the previous term by 2 and adding 1.

Therefore, the sixth term = (2 x 3999) + 1 = 7999.

|||||||||

{ 644 }

Solve this: what comes next in this sequence?:

240, 257, 274, 291, 308, ...?

||||

By taking the differences between adjacent terms, we can see that this is

an Arithmetic Progression, with common difference = 17. Therefore, the

sixth term can be calculated by adding the common difference to the fifth

term. It is equal to 308 + 17 = 325.

|||||||||

{ 645 }

Which number fits in the next spot in this sequence?:

246, 493, 987, 1975, 3951, ...?

||||

By observation and trial and error, we find that each term in this

sequence is obtained by multiplying the previous term by 2 and adding 1.

Therefore, the sixth term = (2 x 3951) + 1 = 7903.

|||||||||

{ 646 }

Which number fits in the next spot in this sequence?:

1.3, 5.2, 20.8, 83.2, 332.8, ...?

||||

By taking the ratios of adjacent terms, we can see that this is a

Geometric Progression, with common ratio = 4. Therefore, the sixth term

can be calculated by multiplying the fifth term with the common ratio. It is

equal to 332.8 x 4 = 1331.2.

|||||||||

{ 647 }

Which number fits in the next spot in this sequence?:

155, 466, 1399, 4198, 12595, ...?

||||

By observation and trial and error, we find that each term in this sequence is obtained by multiplying the previous term by 3 and adding 1. Therefore, the sixth term = (3 x 12595) + 1 = 37786.

||||||||

{ 648 }

Solve this: what comes next in this sequence?:

285, 314, 343, 372, 401, ...?

||||

By taking the differences between adjacent terms, we can see that this is an Arithmetic Progression, with common difference = 29. Therefore, the sixth term can be calculated by adding the common difference to the fifth term. It is equal to 401 + 29 = 430.

|||||||||

{ 649 }

Solve this: what comes next in this sequence?:

194, 240, 286, 332, 378, ...?

||||

By taking the differences between adjacent terms, we can see that this is an Arithmetic Progression, with common difference = 46. Therefore, the sixth term can be calculated by adding the common difference to the fifth term. It is equal to 378 + 46 = 424.

||||||||

{ 650 }

Which number fits in the next spot in this sequence?:

197, 592, 1777, 5332, 15997, ...?

||||

By observation and trial and error, we find that each term in this

sequence is obtained by multiplying the previous term by 3 and adding 1.

Therefore, the sixth term = (3 x 15997) + 1 = 47992.

|||||||||

{ 651 }

Which number fits in the next spot in this sequence?:

8.6, 34.4, 137.6, 550.4, 2201.6, ...?

||||

By taking the ratios of adjacent terms, we can see that this is a

Geometric Progression, with common ratio = 4. Therefore, the sixth term

can be calculated by multiplying the fifth term with the common ratio. It is

equal to 2201.6 x 4 = 8806.4.

|||||||||

{ 652 }

Which number fits in the next spot in this sequence?:

299, 599, 1199, 2399, 4799, ...?

||||

By observation and trial and error, we find that each term in this

sequence is obtained by multiplying the previous term by 2 and adding 1.

Therefore, the sixth term = (2 x 4799) + 1 = 9599.

|||||||||

{ 653 }

Which number fits in the next spot in this sequence?:

226, 905, 3621, 14485, 57941, ...?

||||

By observation and trial and error, we find that each term in this

sequence is obtained by multiplying the previous term by 4 and adding 1.

Therefore, the sixth term = (4 x 57941) + 1 = 231765.

||||||||

{ 654 }

Which number fits in the next spot in this sequence?:

223, 447, 895, 1791, 3583, ...?

||||

By observation and trial and error, we find that each term in this

sequence is obtained by multiplying the previous term by 2 and adding 1.

Therefore, the sixth term = (2 x 3583) + 1 = 7167.

|||||||||

{ 655 }

Which number fits in the next spot in this sequence?:

4.1, 8.2, 16.4, 32.8, 65.6, ...?

||||

By taking the ratios of adjacent terms, we can see that this is a

Geometric Progression, with common ratio = 2. Therefore, the sixth term

can be calculated by multiplying the fifth term with the common ratio. It is

equal to 65.6 x 2 = 131.2.

|||||||||

{ 656 }

Solve this: what comes next in this sequence?:

300, 322, 344, 366, 388, ...?

||||

By taking the differences between adjacent terms, we can see that this is

an Arithmetic Progression, with common difference = 22. Therefore, the

sixth term can be calculated by adding the common difference to the fifth

term. It is equal to 388 + 22 = 410.

|||||||||

{ 657 }

Which number fits in the next spot in this sequence?:

289, 1157, 4629, 18517, 74069, ...?

||||

By observation and trial and error, we find that each term in this

sequence is obtained by multiplying the previous term by 4 and adding 1.

Therefore, the sixth term = (4 x 74069) + 1 = 296277.

|||||||||

{ 658 }

Which number fits in the next spot in this sequence?:

5.5, 16.5, 49.5, 148.5, 445.5, ...?

||||

By taking the ratios of adjacent terms, we can see that this is a

Geometric Progression, with common ratio = 3. Therefore, the sixth term

can be calculated by multiplying the fifth term with the common ratio. It is

equal to 445.5 x 3 = 1336.5.

|||||||||

{ 659 }

Which number fits in the next spot in this sequence?:

98, 393, 1573, 6293, 25173, ...?

||||

By observation and trial and error, we find that each term in this sequence is obtained by multiplying the previous term by 4 and adding 1. Therefore, the sixth term = (4 x 25173) + 1 = 100693.

|||||||||

{ 660 }

Which number fits in the next spot in this sequence?:

7.2, 21.6, 64.8, 194.4, 583.2, ...?

||||

By taking the ratios of adjacent terms, we can see that this is a Geometric Progression, with common ratio = 3. Therefore, the sixth term can be calculated by multiplying the fifth term with the common ratio. It is equal to 583.2 x 3 = 1749.6.

|||||||||

{ 661 }

Which number fits in the next spot in this sequence?:

4.8, 19.2, 76.8, 307.2, 1228.8, ...?

||||

By taking the ratios of adjacent terms, we can see that this is a Geometric Progression, with common ratio = 4. Therefore, the sixth term can be calculated by multiplying the fifth term with the common ratio. It is equal to 1228.8 x 4 = 4915.2.

||||||||

{ 662 }

Which number fits in the next spot in this sequence?:

7.9, 31.6, 126.4, 505.6, 2022.4, ...?

||||

By taking the ratios of adjacent terms, we can see that this is a

Geometric Progression, with common ratio = 4. Therefore, the sixth term

can be calculated by multiplying the fifth term with the common ratio. It is

equal to 2022.4 x 4 = 8089.6.

||||||||

{ 663 }

Which number fits in the next spot in this sequence?:

3.8, 15.2, 60.8, 243.2, 972.8, ...?

||||

By taking the ratios of adjacent terms, we can see that this is a

Geometric Progression, with common ratio = 4. Therefore, the sixth term

can be calculated by multiplying the fifth term with the common ratio. It is

equal to 972.8 x 4 = 3891.2.

||||||||

{ 664 }

Solve this: what comes next in this sequence?:

263, 311, 359, 407, 455, ...?

||||

By taking the differences between adjacent terms, we can see that this is an Arithmetic Progression, with common difference = 48. Therefore, the sixth term can be calculated by adding the common difference to the fifth term. It is equal to 455 + 48 = 503.

|||||||||

{ 665 }

Which number fits in the next spot in this sequence?:

6.9, 13.8, 27.6, 55.2, 110.4, ...?

||||

By taking the ratios of adjacent terms, we can see that this is a Geometric Progression, with common ratio = 2. Therefore, the sixth term can be calculated by multiplying the fifth term with the common ratio. It is equal to 110.4 x 2 = 220.8.

|||||||||

{ 666 }

Solve this: what comes next in this sequence?:

117, 163, 209, 255, 301, ...?

||||

By taking the differences between adjacent terms, we can see that this is

an Arithmetic Progression, with common difference = 46. Therefore, the

sixth term can be calculated by adding the common difference to the fifth

term. It is equal to 301 + 46 = 347.

|||||||||

{ 667 }

Which number fits in the next spot in this sequence?:

213, 427, 855, 1711, 3423, ...?

||||

By observation and trial and error, we find that each term in this

sequence is obtained by multiplying the previous term by 2 and adding 1.

Therefore, the sixth term = (2 x 3423) + 1 = 6847.

|||||||||

{ 668 }

Which number fits in the next spot in this sequence?:

199, 598, 1795, 5386, 16159, ...?

||||

By observation and trial and error, we find that each term in this

sequence is obtained by multiplying the previous term by 3 and adding 1.

Therefore, the sixth term = (3 x 16159) + 1 = 48478.

IIIIIIII

{ 669 }

Solve this: what comes next in this sequence?:

230, 241, 252, 263, 274, ...?

IIII

By taking the differences between adjacent terms, we can see that this is

an Arithmetic Progression, with common difference = 11. Therefore, the

sixth term can be calculated by adding the common difference to the fifth

term. It is equal to 274 + 11 = 285.

IIIIIIII

{ 670 }

Solve this: what comes next in this sequence?:

120, 149, 178, 207, 236, ...?

IIII

By taking the differences between adjacent terms, we can see that this is

an Arithmetic Progression, with common difference = 29. Therefore, the

sixth term can be calculated by adding the common difference to the fifth

term. It is equal to 236 + 29 = 265.

IIIIIIII

{ 671 }

Which number fits in the next spot in this sequence?:

206, 825, 3301, 13205, 52821, ...?

||||

By observation and trial and error, we find that each term in this

sequence is obtained by multiplying the previous term by 4 and adding 1.

Therefore, the sixth term = (4 x 52821) + 1 = 211285.

|||||||||

{ 672 }

Which number fits in the next spot in this sequence?:

2.4, 7.2, 21.6, 64.8, 194.4, ...?

||||

By taking the ratios of adjacent terms, we can see that this is a

Geometric Progression, with common ratio = 3. Therefore, the sixth term

can be calculated by multiplying the fifth term with the common ratio. It is

equal to 194.4 x 3 = 583.2.

|||||||||

{ 673 }

Solve this: what comes next in this sequence?:

193, 243, 293, 343, 393, ...?

||||

By taking the differences between adjacent terms, we can see that this is an Arithmetic Progression, with common difference = 50. Therefore, the sixth term can be calculated by adding the common difference to the fifth term. It is equal to 393 + 50 = 443.

||||||||

{ 674 }

Which number fits in the next spot in this sequence?:

9.5, 38, 152, 608, 2432, ...?

||||

By taking the ratios of adjacent terms, we can see that this is a Geometric Progression, with common ratio = 4. Therefore, the sixth term can be calculated by multiplying the fifth term with the common ratio. It is equal to 2432 x 4 = 9728.

||||||||

{ 675 }

Solve this: what comes next in this sequence?:

157, 185, 213, 241, 269, ...?

||||

By taking the differences between adjacent terms, we can see that this is an Arithmetic Progression, with common difference = 28. Therefore, the sixth term can be calculated by adding the common difference to the fifth

term. It is equal to 269 + 28 = 297.

|||||||||

{ 676 }

Which number fits in the next spot in this sequence?:

3.6, 10.8, 32.4, 97.2, 291.6, ...?

||||

By taking the ratios of adjacent terms, we can see that this is a

Geometric Progression, with common ratio = 3. Therefore, the sixth term

can be calculated by multiplying the fifth term with the common ratio. It is

equal to 291.6 x 3 = 874.800000000001.

|||||||||

{ 677 }

Which number fits in the next spot in this sequence?:

8.7, 34.8, 139.2, 556.8, 2227.2, ...?

||||

By taking the ratios of adjacent terms, we can see that this is a

Geometric Progression, with common ratio = 4. Therefore, the sixth term

can be calculated by multiplying the fifth term with the common ratio. It is

equal to 2227.2 x 4 = 8908.8.

|||||||||

{ 678 }

Solve this: what comes next in this sequence?:

227, 237, 247, 257, 267, ...?

||||

By taking the differences between adjacent terms, we can see that this is an Arithmetic Progression, with common difference = 10. Therefore, the sixth term can be calculated by adding the common difference to the fifth term. It is equal to 267 + 10 = 277.

|||||||||

{ 679 }

Solve this: what comes next in this sequence?:

105, 124, 143, 162, 181, ...?

||||

By taking the differences between adjacent terms, we can see that this is an Arithmetic Progression, with common difference = 19. Therefore, the sixth term can be calculated by adding the common difference to the fifth term. It is equal to 181 + 19 = 200.

|||||||||

{ 680 }

Which number fits in the next spot in this sequence?:

259, 519, 1039, 2079, 4159, ...?

||||

By observation and trial and error, we find that each term in this sequence is obtained by multiplying the previous term by 2 and adding 1. Therefore, the sixth term = (2 x 4159) + 1 = 8319.

|||||||||

{ 681 }

Which number fits in the next spot in this sequence?:

8.1, 16.2, 32.4, 64.8, 129.6, ...?

||||

By taking the ratios of adjacent terms, we can see that this is a Geometric Progression, with common ratio = 2. Therefore, the sixth term can be calculated by multiplying the fifth term with the common ratio. It is equal to 129.6 x 2 = 259.2.

|||||||||

{ 682 }

Solve this: what comes next in this sequence?:

197, 215, 233, 251, 269, ...?

||||

By taking the differences between adjacent terms, we can see that this is an Arithmetic Progression, with common difference = 18. Therefore, the sixth term can be calculated by adding the common difference to the fifth

term. It is equal to 269 + 18 = 287.

|||||||||

{ 683 }

Which number fits in the next spot in this sequence?:

9.1, 27.3, 81.9, 245.7, 737.1, ...?

||||

By taking the ratios of adjacent terms, we can see that this is a

Geometric Progression, with common ratio = 3. Therefore, the sixth term

can be calculated by multiplying the fifth term with the common ratio. It is

equal to 737.1 x 3 = 2211.3.

|||||||||

{ 684 }

Which number fits in the next spot in this sequence?:

4.4, 13.2, 39.6, 118.8, 356.4, ...?

||||

By taking the ratios of adjacent terms, we can see that this is a

Geometric Progression, with common ratio = 3. Therefore, the sixth term

can be calculated by multiplying the fifth term with the common ratio. It is

equal to 356.4 x 3 = 1069.2.

|||||||||

{ 685 }

Which number fits in the next spot in this sequence?:

220, 441, 883, 1767, 3535, ...?

||||

By observation and trial and error, we find that each term in this sequence is obtained by multiplying the previous term by 2 and adding 1. Therefore, the sixth term = (2 x 3535) + 1 = 7071.

|||||||||

{ 686 }

Which number fits in the next spot in this sequence?:

4.6, 18.4, 73.6, 294.4, 1177.6, ...?

||||

By taking the ratios of adjacent terms, we can see that this is a Geometric Progression, with common ratio = 4. Therefore, the sixth term can be calculated by multiplying the fifth term with the common ratio. It is equal to 1177.6 x 4 = 4710.4.

||||||||

{ 687 }

Which number fits in the next spot in this sequence?:

212, 425, 851, 1703, 3407, ...?

||||

By observation and trial and error, we find that each term in this sequence is obtained by multiplying the previous term by 2 and adding 1. Therefore, the sixth term = (2 x 3407) + 1 = 6815.

|||||||||

{ 688 }

Which number fits in the next spot in this sequence?:

115, 346, 1039, 3118, 9355, ...?

||||

By observation and trial and error, we find that each term in this sequence is obtained by multiplying the previous term by 3 and adding 1. Therefore, the sixth term = (3 x 9355) + 1 = 28066.

|||||||||

{ 689 }

Solve this: what comes next in this sequence?:

131, 181, 231, 281, 331, ...?

||||

By taking the differences between adjacent terms, we can see that this is an Arithmetic Progression, with common difference = 50. Therefore, the sixth term can be calculated by adding the common difference to the fifth term. It is equal to 331 + 50 = 381.

|||||||||

{ 690 }

Solve this: what comes next in this sequence?:

185, 211, 237, 263, 289, ...?

||||

By taking the differences between adjacent terms, we can see that this is an Arithmetic Progression, with common difference = 26. Therefore, the sixth term can be calculated by adding the common difference to the fifth term. It is equal to 289 + 26 = 315.

|||||||||

{ 691 }

Solve this: what comes next in this sequence?:

167, 185, 203, 221, 239, ...?

||||

By taking the differences between adjacent terms, we can see that this is an Arithmetic Progression, with common difference = 18. Therefore, the sixth term can be calculated by adding the common difference to the fifth term. It is equal to 239 + 18 = 257.

|||||||||

{ 692 }

Which number fits in the next spot in this sequence?:

250, 501, 1003, 2007, 4015, ...?

||||

By observation and trial and error, we find that each term in this sequence is obtained by multiplying the previous term by 2 and adding 1. Therefore, the sixth term = (2 x 4015) + 1 = 8031.

|||||||||

{ 693 }

Which number fits in the next spot in this sequence?:

6.3, 12.6, 25.2, 50.4, 100.8, ...?

||||

By taking the ratios of adjacent terms, we can see that this is a Geometric Progression, with common ratio = 2. Therefore, the sixth term can be calculated by multiplying the fifth term with the common ratio. It is equal to 100.8 x 2 = 201.6.

|||||||||

{ 694 }

Solve this: what comes next in this sequence?:

201, 232, 263, 294, 325, ...?

||||

By taking the differences between adjacent terms, we can see that this is an Arithmetic Progression, with common difference = 31. Therefore, the sixth term can be calculated by adding the common difference to the fifth

term. It is equal to 325 + 31 = 356.

|||||||||

{ 695 }

Which number fits in the next spot in this sequence?:

234, 469, 939, 1879, 3759, ...?

||||

By observation and trial and error, we find that each term in this

sequence is obtained by multiplying the previous term by 2 and adding 1.

Therefore, the sixth term = (2 x 3759) + 1 = 7519.

|||||||||

{ 696 }

Which number fits in the next spot in this sequence?:

7, 14, 28, 56, 112, ...?

||||

By taking the ratios of adjacent terms, we can see that this is a

Geometric Progression, with common ratio = 2. Therefore, the sixth term

can be calculated by multiplying the fifth term with the common ratio. It is

equal to 112 x 2 = 224.

|||||||||

{ 697 }

Which number fits in the next spot in this sequence?:

5.4, 21.6, 86.4, 345.6, 1382.4, ...?

||||

By taking the ratios of adjacent terms, we can see that this is a

Geometric Progression, with common ratio = 4. Therefore, the sixth term

can be calculated by multiplying the fifth term with the common ratio. It is

equal to 1382.4 x 4 = 5529.6.

|||||||||

{ 698 }

Which number fits in the next spot in this sequence?:

9.3, 18.6, 37.2, 74.4, 148.8, ...?

||||

By taking the ratios of adjacent terms, we can see that this is a

Geometric Progression, with common ratio = 2. Therefore, the sixth term

can be calculated by multiplying the fifth term with the common ratio. It is

equal to 148.8 x 2 = 297.6.

|||||||||

{ 699 }

Solve this: what comes next in this sequence?:

156, 197, 238, 279, 320, ...?

||||

By taking the differences between adjacent terms, we can see that this is

an Arithmetic Progression, with common difference = 41. Therefore, the

sixth term can be calculated by adding the common difference to the fifth

term. It is equal to 320 + 41 = 361.

|||||||||

{ 700 }

Which number fits in the next spot in this sequence?:

149, 448, 1345, 4036, 12109, ...?

||||

By observation and trial and error, we find that each term in this

sequence is obtained by multiplying the previous term by 3 and adding 1.

Therefore, the sixth term = (3 x 12109) + 1 = 36328.

|||||||||

{ 701 }

Which number fits in the next spot in this sequence?:

7.5, 15, 30, 60, 120, ...?

||||

By taking the ratios of adjacent terms, we can see that this is a

Geometric Progression, with common ratio = 2. Therefore, the sixth term

can be calculated by multiplying the fifth term with the common ratio. It is

equal to 120 x 2 = 240.

||||||||

{ 702 }

Solve this: what comes next in this sequence?:

147, 161, 175, 189, 203, ...?

||||

By taking the differences between adjacent terms, we can see that this is

an Arithmetic Progression, with common difference = 14. Therefore, the

sixth term can be calculated by adding the common difference to the fifth

term. It is equal to 203 + 14 = 217.

||||||||

{ 703 }

Solve this: what comes next in this sequence?:

190, 224, 258, 292, 326, ...?

||||

By taking the differences between adjacent terms, we can see that this is

an Arithmetic Progression, with common difference = 34. Therefore, the

sixth term can be calculated by adding the common difference to the fifth

term. It is equal to 326 + 34 = 360.

||||||||

{ 704 }

Which number fits in the next spot in this sequence?:

162, 487, 1462, 4387, 13162, ...?

||||

By observation and trial and error, we find that each term in this sequence is obtained by multiplying the previous term by 3 and adding 1. Therefore, the sixth term = (3 x 13162) + 1 = 39487.

|||||||||

{ 705 }

Solve this: what comes next in this sequence?:

116, 153, 190, 227, 264, ...?

||||

By taking the differences between adjacent terms, we can see that this is an Arithmetic Progression, with common difference = 37. Therefore, the sixth term can be calculated by adding the common difference to the fifth term. It is equal to 264 + 37 = 301.

|||||||||

{ 706 }

Which number fits in the next spot in this sequence?:

109, 219, 439, 879, 1759, ...?

||||

By observation and trial and error, we find that each term in this sequence is obtained by multiplying the previous term by 2 and adding 1. Therefore, the sixth term = (2 x 1759) + 1 = 3519.

||||||||

{ 707 }

Which number fits in the next spot in this sequence?:

7.7, 30.8, 123.2, 492.8, 1971.2, ...?

||||

By taking the ratios of adjacent terms, we can see that this is a Geometric Progression, with common ratio = 4. Therefore, the sixth term can be calculated by multiplying the fifth term with the common ratio. It is equal to 1971.2 x 4 = 7884.8.

||||||||

{ 708 }

Solve this: what comes next in this sequence?:

181, 194, 207, 220, 233, ...?

||||

By taking the differences between adjacent terms, we can see that this is an Arithmetic Progression, with common difference = 13. Therefore, the sixth term can be calculated by adding the common difference to the fifth term. It is equal to 233 + 13 = 246.

||||||||

{ 709 }

Solve this: what comes next in this sequence?:

240, 280, 320, 360, 400, ...?

||||

By taking the differences between adjacent terms, we can see that this is

an Arithmetic Progression, with common difference = 40. Therefore, the

sixth term can be calculated by adding the common difference to the fifth

term. It is equal to 400 + 40 = 440.

||||||||

{ 710 }

Which number fits in the next spot in this sequence?:

292, 877, 2632, 7897, 23692, ...?

||||

By observation and trial and error, we find that each term in this

sequence is obtained by multiplying the previous term by 3 and adding 1.

Therefore, the sixth term = (3 x 23692) + 1 = 71077.

||||||||

{ 711 }

Which number fits in the next spot in this sequence?:

280, 561, 1123, 2247, 4495, ...?

||||

By observation and trial and error, we find that each term in this

sequence is obtained by multiplying the previous term by 2 and adding 1.

Therefore, the sixth term = (2 x 4495) + 1 = 8991.

|||||||||

{ 712 }

Solve this: what comes next in this sequence?:

159, 207, 255, 303, 351, ...?

||||

By taking the differences between adjacent terms, we can see that this is

an Arithmetic Progression, with common difference = 48. Therefore, the

sixth term can be calculated by adding the common difference to the fifth

term. It is equal to 351 + 48 = 399.

|||||||||

{ 713 }

Which number fits in the next spot in this sequence?:

211, 423, 847, 1695, 3391, ...?

||||

By observation and trial and error, we find that each term in this

sequence is obtained by multiplying the previous term by 2 and adding 1.

Therefore, the sixth term = (2 x 3391) + 1 = 6783.

||||||||

{ 714 }

Which number fits in the next spot in this sequence?:

261, 523, 1047, 2095, 4191, ...?

||||

By observation and trial and error, we find that each term in this

sequence is obtained by multiplying the previous term by 2 and adding 1.

Therefore, the sixth term = (2 x 4191) + 1 = 8383.

||||||||

{ 715 }

Which number fits in the next spot in this sequence?:

159, 637, 2549, 10197, 40789, ...?

||||

By observation and trial and error, we find that each term in this

sequence is obtained by multiplying the previous term by 4 and adding 1.

Therefore, the sixth term = (4 x 40789) + 1 = 163157.

||||||||

{ 716 }

Solve this: what comes next in this sequence?:

140, 157, 174, 191, 208, ...?

||||

By taking the differences between adjacent terms, we can see that this is

an Arithmetic Progression, with common difference = 17. Therefore, the

sixth term can be calculated by adding the common difference to the fifth

term. It is equal to 208 + 17 = 225.

|||||||||

{ 717 }

Solve this: what comes next in this sequence?:

90, 114, 138, 162, 186, ...?

||||

By taking the differences between adjacent terms, we can see that this is

an Arithmetic Progression, with common difference = 24. Therefore, the

sixth term can be calculated by adding the common difference to the fifth

term. It is equal to 186 + 24 = 210.

|||||||||

{ 718 }

Which number fits in the next spot in this sequence?:

2.3, 9.2, 36.8, 147.2, 588.8, ...?

||||

By taking the ratios of adjacent terms, we can see that this is a Geometric Progression, with common ratio = 4. Therefore, the sixth term can be calculated by multiplying the fifth term with the common ratio. It is equal to 588.8 x 4 = 2355.2.

|||||||||

{ 719 }

Solve this: what comes next in this sequence?:

221, 245, 269, 293, 317, ...?

||||

By taking the differences between adjacent terms, we can see that this is an Arithmetic Progression, with common difference = 24. Therefore, the sixth term can be calculated by adding the common difference to the fifth term. It is equal to 317 + 24 = 341.

||||||||

{ 720 }

Which number fits in the next spot in this sequence?:

3.3, 6.6, 13.2, 26.4, 52.8, ...?

||||

By taking the ratios of adjacent terms, we can see that this is a Geometric Progression, with common ratio = 2. Therefore, the sixth term

can be calculated by multiplying the fifth term with the common ratio. It is equal to 52.8 x 2 = 105.6.

||||||||||

{ 721 }

Solve this: what comes next in this sequence?:

242, 267, 292, 317, 342, ...?

|||||

By taking the differences between adjacent terms, we can see that this is an Arithmetic Progression, with common difference = 25. Therefore, the sixth term can be calculated by adding the common difference to the fifth term. It is equal to 342 + 25 = 367.

||||||||||

{ 722 }

Which number fits in the next spot in this sequence?:

2.9, 5.8, 11.6, 23.2, 46.4, ...?

|||||

By taking the ratios of adjacent terms, we can see that this is a Geometric Progression, with common ratio = 2. Therefore, the sixth term can be calculated by multiplying the fifth term with the common ratio. It is equal to 46.4 x 2 = 92.8000000000001.

||||||||||

{ 723 }

Which number fits in the next spot in this sequence?:

280, 1121, 4485, 17941, 71765, ...?

||||

By observation and trial and error, we find that each term in this

sequence is obtained by multiplying the previous term by 4 and adding 1.

Therefore, the sixth term = (4 x 71765) + 1 = 287061.

|||||||||

{ 724 }

Which number fits in the next spot in this sequence?:

260, 1041, 4165, 16661, 66645, ...?

||||

By observation and trial and error, we find that each term in this

sequence is obtained by multiplying the previous term by 4 and adding 1.

Therefore, the sixth term = (4 x 66645) + 1 = 266581.

|||||||||

{ 725 }

Solve this: what comes next in this sequence?:

134, 163, 192, 221, 250, ...?

||||

By taking the differences between adjacent terms, we can see that this is

an Arithmetic Progression, with common difference = 29. Therefore, the sixth term can be calculated by adding the common difference to the fifth term. It is equal to 250 + 29 = 279.

IIIIIIII

{ 726 }

Which number fits in the next spot in this sequence?:

9.9, 19.8, 39.6, 79.2, 158.4, ...?

IIII

By taking the ratios of adjacent terms, we can see that this is a Geometric Progression, with common ratio = 2. Therefore, the sixth term can be calculated by multiplying the fifth term with the common ratio. It is equal to 158.4 x 2 = 316.8.

IIIIIIIII

{ 727 }

Solve this: what comes next in this sequence?:

160, 193, 226, 259, 292, ...?

IIII

By taking the differences between adjacent terms, we can see that this is an Arithmetic Progression, with common difference = 33. Therefore, the sixth term can be calculated by adding the common difference to the fifth term. It is equal to 292 + 33 = 325.

IIIIIIII

{ 728 }

Which number fits in the next spot in this sequence?:

6.7, 13.4, 26.8, 53.6, 107.2, ...?

IIII

By taking the ratios of adjacent terms, we can see that this is a

Geometric Progression, with common ratio = 2. Therefore, the sixth term

can be calculated by multiplying the fifth term with the common ratio. It is

equal to 107.2 x 2 = 214.4.

IIIIIIII

{ 729 }

Which number fits in the next spot in this sequence?:

7, 21, 63, 189, 567, ...?

IIII

By taking the ratios of adjacent terms, we can see that this is a

Geometric Progression, with common ratio = 3. Therefore, the sixth term

can be calculated by multiplying the fifth term with the common ratio. It is

equal to 567 x 3 = 1701.

IIIIIIII

{ 730 }

Which number fits in the next spot in this sequence?:

1.2, 4.8, 19.2, 76.8, 307.2, ...?

||||

By taking the ratios of adjacent terms, we can see that this is a

Geometric Progression, with common ratio = 4. Therefore, the sixth term

can be calculated by multiplying the fifth term with the common ratio. It is

equal to 307.2 x 4 = 1228.8.

|||||||||

{ 731 }

Solve this: what comes next in this sequence?:

118, 141, 164, 187, 210, ...?

||||

By taking the differences between adjacent terms, we can see that this is

an Arithmetic Progression, with common difference = 23. Therefore, the

sixth term can be calculated by adding the common difference to the fifth

term. It is equal to 210 + 23 = 233.

|||||||||

{ 732 }

Which number fits in the next spot in this sequence?:

6.7, 20.1, 60.3, 180.9, 542.7, ...?

||||

By taking the ratios of adjacent terms, we can see that this is a Geometric Progression, with common ratio = 3. Therefore, the sixth term can be calculated by multiplying the fifth term with the common ratio. It is equal to 542.7 x 3 = 1628.1.

|||||||||

{ 733 }

Solve this: what comes next in this sequence?:

143, 154, 165, 176, 187, ...?

||||

By taking the differences between adjacent terms, we can see that this is an Arithmetic Progression, with common difference = 11. Therefore, the sixth term can be calculated by adding the common difference to the fifth term. It is equal to 187 + 11 = 198.

|||||||||

{ 734 }

Which number fits in the next spot in this sequence?:

224, 897, 3589, 14357, 57429, ...?

||||

By observation and trial and error, we find that each term in this sequence is obtained by multiplying the previous term by 4 and adding 1.

Therefore, the sixth term = (4 x 57429) + 1 = 229717.

|||||||||

{ 735 }

Which number fits in the next spot in this sequence?:

7.1, 28.4, 113.6, 454.4, 1817.6, ...?

||||

By taking the ratios of adjacent terms, we can see that this is a

Geometric Progression, with common ratio = 4. Therefore, the sixth term

can be calculated by multiplying the fifth term with the common ratio. It is

equal to 1817.6 x 4 = 7270.4.

|||||||||

{ 736 }

Which number fits in the next spot in this sequence?:

6.3, 18.9, 56.7, 170.1, 510.3, ...?

||||

By taking the ratios of adjacent terms, we can see that this is a

Geometric Progression, with common ratio = 3. Therefore, the sixth term

can be calculated by multiplying the fifth term with the common ratio. It is

equal to 510.3 x 3 = 1530.9.

|||||||||

{ 737 }

Which number fits in the next spot in this sequence?:

293, 880, 2641, 7924, 23773, ...?

||||

By observation and trial and error, we find that each term in this

sequence is obtained by multiplying the previous term by 3 and adding 1.

Therefore, the sixth term = (3 x 23773) + 1 = 71320.

|||||||||

{ 738 }

Which number fits in the next spot in this sequence?:

117, 469, 1877, 7509, 30037, ...?

||||

By observation and trial and error, we find that each term in this

sequence is obtained by multiplying the previous term by 4 and adding 1.

Therefore, the sixth term = (4 x 30037) + 1 = 120149.

|||||||||

{ 739 }

Which number fits in the next spot in this sequence?:

8.4, 16.8, 33.6, 67.2, 134.4, ...?

||||

By taking the ratios of adjacent terms, we can see that this is a

Geometric Progression, with common ratio = 2. Therefore, the sixth term can be calculated by multiplying the fifth term with the common ratio. It is equal to 134.4 x 2 = 268.8.

||||||||

{ 740 }

Which number fits in the next spot in this sequence?:

266, 1065, 4261, 17045, 68181, ...?

||||

By observation and trial and error, we find that each term in this sequence is obtained by multiplying the previous term by 4 and adding 1. Therefore, the sixth term = (4 x 68181) + 1 = 272725.

||||||||

{ 741 }

Solve this: what comes next in this sequence?:

180, 196, 212, 228, 244, ...?

||||

By taking the differences between adjacent terms, we can see that this is an Arithmetic Progression, with common difference = 16. Therefore, the sixth term can be calculated by adding the common difference to the fifth term. It is equal to 244 + 16 = 260.

||||||||

{ 742 }

Which number fits in the next spot in this sequence?:

116, 465, 1861, 7445, 29781, ...?

||||

By observation and trial and error, we find that each term in this sequence is obtained by multiplying the previous term by 4 and adding 1. Therefore, the sixth term = (4 x 29781) + 1 = 119125.

|||||||||

{ 743 }

Solve this: what comes next in this sequence?:

144, 192, 240, 288, 336, ...?

||||

By taking the differences between adjacent terms, we can see that this is an Arithmetic Progression, with common difference = 48. Therefore, the sixth term can be calculated by adding the common difference to the fifth term. It is equal to 336 + 48 = 384.

|||||||||

{ 744 }

Which number fits in the next spot in this sequence?:

3.3, 6.6, 13.2, 26.4, 52.8, ...?

||||

By taking the ratios of adjacent terms, we can see that this is a

Geometric Progression, with common ratio = 2. Therefore, the sixth term

can be calculated by multiplying the fifth term with the common ratio. It is

equal to 52.8 x 2 = 105.6.

||||||||

{ 745 }

Solve this: what comes next in this sequence?:

297, 308, 319, 330, 341, ...?

||||

By taking the differences between adjacent terms, we can see that this is

an Arithmetic Progression, with common difference = 11. Therefore, the

sixth term can be calculated by adding the common difference to the fifth

term. It is equal to 341 + 11 = 352.

||||||||

{ 746 }

Which number fits in the next spot in this sequence?:

132, 397, 1192, 3577, 10732, ...?

||||

By observation and trial and error, we find that each term in this

sequence is obtained by multiplying the previous term by 3 and adding 1.

Therefore, the sixth term = (3 x 10732) + 1 = 32197.

IIIIIIII

{ 747 }

Which number fits in the next spot in this sequence?:

140, 421, 1264, 3793, 11380, ...?

IIII

By observation and trial and error, we find that each term in this

sequence is obtained by multiplying the previous term by 3 and adding 1.

Therefore, the sixth term = (3 x 11380) + 1 = 34141.

IIIIIIII

{ 748 }

Which number fits in the next spot in this sequence?:

8.3, 16.6, 33.2, 66.4, 132.8, ...?

IIII

By taking the ratios of adjacent terms, we can see that this is a

Geometric Progression, with common ratio = 2. Therefore, the sixth term

can be calculated by multiplying the fifth term with the common ratio. It is

equal to 132.8 x 2 = 265.6.

IIIIIIII

{ 749 }

Which number fits in the next spot in this sequence?:

122, 245, 491, 983, 1967, ...?

||||

By observation and trial and error, we find that each term in this

sequence is obtained by multiplying the previous term by 2 and adding 1.

Therefore, the sixth term = (2 x 1967) + 1 = 3935.

|||||||||

{ 750 }

Which number fits in the next spot in this sequence?:

1.3, 3.9, 11.7, 35.1, 105.3, ...?

||||

By taking the ratios of adjacent terms, we can see that this is a

Geometric Progression, with common ratio = 3. Therefore, the sixth term

can be calculated by multiplying the fifth term with the common ratio. It is

equal to 105.3 x 3 = 315.9.

|||||||||

{ 751 }

Solve this: what comes next in this sequence?:

174, 189, 204, 219, 234, ...?

||||

By taking the differences between adjacent terms, we can see that this is an Arithmetic Progression, with common difference = 15. Therefore, the sixth term can be calculated by adding the common difference to the fifth term. It is equal to 234 + 15 = 249.

||||||||

{ 752 }

Which number fits in the next spot in this sequence?:

3.8, 11.4, 34.2, 102.6, 307.8, ...?

||||

By taking the ratios of adjacent terms, we can see that this is a Geometric Progression, with common ratio = 3. Therefore, the sixth term can be calculated by multiplying the fifth term with the common ratio. It is equal to 307.8 x 3 = 923.4.

||||||||

{ 753 }

Which number fits in the next spot in this sequence?:

9.7, 38.8, 155.2, 620.8, 2483.2, ...?

||||

By taking the ratios of adjacent terms, we can see that this is a Geometric Progression, with common ratio = 4. Therefore, the sixth term can be calculated by multiplying the fifth term with the common ratio. It is

equal to 2483.2 x 4 = 9932.8.

||||||||

{ 754 }

Which number fits in the next spot in this sequence?:

223, 893, 3573, 14293, 57173, ...?

||||

By observation and trial and error, we find that each term in this

sequence is obtained by multiplying the previous term by 4 and adding 1.

Therefore, the sixth term = (4 x 57173) + 1 = 228693.

|||||||||

{ 755 }

Which number fits in the next spot in this sequence?:

213, 853, 3413, 13653, 54613, ...?

||||

By observation and trial and error, we find that each term in this

sequence is obtained by multiplying the previous term by 4 and adding 1.

Therefore, the sixth term = (4 x 54613) + 1 = 218453.

|||||||||

{ 756 }

Solve this: what comes next in this sequence?:

286, 322, 358, 394, 430, ...?

||||

By taking the differences between adjacent terms, we can see that this is

an Arithmetic Progression, with common difference = 36. Therefore, the

sixth term can be calculated by adding the common difference to the fifth

term. It is equal to 430 + 36 = 466.

|||||||||

{ 757 }

Solve this: what comes next in this sequence?:

124, 171, 218, 265, 312, ...?

||||

By taking the differences between adjacent terms, we can see that this is

an Arithmetic Progression, with common difference = 47. Therefore, the

sixth term can be calculated by adding the common difference to the fifth

term. It is equal to 312 + 47 = 359.

|||||||||

{ 758 }

Which number fits in the next spot in this sequence?:

115, 346, 1039, 3118, 9355, ...?

IIII

By observation and trial and error, we find that each term in this sequence is obtained by multiplying the previous term by 3 and adding 1. Therefore, the sixth term = (3 x 9355) + 1 = 28066.

IIIIIIIII

{ 759 }

Which number fits in the next spot in this sequence?:

220, 881, 3525, 14101, 56405, ...?

IIII

By observation and trial and error, we find that each term in this sequence is obtained by multiplying the previous term by 4 and adding 1. Therefore, the sixth term = (4 x 56405) + 1 = 225621.

IIIIIIIII

{ 760 }

Solve this: what comes next in this sequence?:

234, 256, 278, 300, 322, ...?

IIII

By taking the differences between adjacent terms, we can see that this is an Arithmetic Progression, with common difference = 22. Therefore, the sixth term can be calculated by adding the common difference to the fifth term. It is equal to 322 + 22 = 344.

|||||||

{ 761 }

Which number fits in the next spot in this sequence?:

285, 856, 2569, 7708, 23125, ...?

||||

By observation and trial and error, we find that each term in this

sequence is obtained by multiplying the previous term by 3 and adding 1.

Therefore, the sixth term = (3 x 23125) + 1 = 69376.

|||||||

{ 762 }

Which number fits in the next spot in this sequence?:

3.5, 14, 56, 224, 896, ...?

||||

By taking the ratios of adjacent terms, we can see that this is a

Geometric Progression, with common ratio = 4. Therefore, the sixth term

can be calculated by multiplying the fifth term with the common ratio. It is

equal to 896 x 4 = 3584.

|||||||

{ 763 }

Which number fits in the next spot in this sequence?:

3.6, 14.4, 57.6, 230.4, 921.6, ...?

||||

By taking the ratios of adjacent terms, we can see that this is a

Geometric Progression, with common ratio = 4. Therefore, the sixth term

can be calculated by multiplying the fifth term with the common ratio. It is

equal to 921.6 x 4 = 3686.4.

|||||||||

{ 764 }

Which number fits in the next spot in this sequence?:

207, 622, 1867, 5602, 16807, ...?

||||

By observation and trial and error, we find that each term in this

sequence is obtained by multiplying the previous term by 3 and adding 1.

Therefore, the sixth term = (3 x 16807) + 1 = 50422.

|||||||||

{ 765 }

Solve this: what comes next in this sequence?:

169, 190, 211, 232, 253, ...?

||||

By taking the differences between adjacent terms, we can see that this is an Arithmetic Progression, with common difference = 21. Therefore, the sixth term can be calculated by adding the common difference to the fifth term. It is equal to 253 + 21 = 274.

IIIIIIII

{ 766 }

Which number fits in the next spot in this sequence?:

192, 385, 771, 1543, 3087, ...?

IIII

By observation and trial and error, we find that each term in this sequence is obtained by multiplying the previous term by 2 and adding 1. Therefore, the sixth term = (2 x 3087) + 1 = 6175.

IIIIIIII

{ 767 }

Solve this: what comes next in this sequence?:

291, 336, 381, 426, 471, ...?

IIII

By taking the differences between adjacent terms, we can see that this is an Arithmetic Progression, with common difference = 45. Therefore, the sixth term can be calculated by adding the common difference to the fifth term. It is equal to 471 + 45 = 516.

||||||||

{ 768 }

Which number fits in the next spot in this sequence?:

191, 383, 767, 1535, 3071, ...?

||||

By observation and trial and error, we find that each term in this

sequence is obtained by multiplying the previous term by 2 and adding 1.

Therefore, the sixth term = (2 x 3071) + 1 = 6143.

|||||||||

{ 769 }

Which number fits in the next spot in this sequence?:

6.3, 18.9, 56.7, 170.1, 510.3, ...?

||||

By taking the ratios of adjacent terms, we can see that this is a

Geometric Progression, with common ratio = 3. Therefore, the sixth term

can be calculated by multiplying the fifth term with the common ratio. It is

equal to 510.3 x 3 = 1530.9.

|||||||||

{ 770 }

Which number fits in the next spot in this sequence?:

4.7, 18.8, 75.2, 300.8, 1203.2, ...?

||||

By taking the ratios of adjacent terms, we can see that this is a

Geometric Progression, with common ratio = 4. Therefore, the sixth term

can be calculated by multiplying the fifth term with the common ratio. It is

equal to 1203.2 x 4 = 4812.8.

||||||||

{ 771 }

Which number fits in the next spot in this sequence?:

233, 933, 3733, 14933, 59733, ...?

||||

By observation and trial and error, we find that each term in this

sequence is obtained by multiplying the previous term by 4 and adding 1.

Therefore, the sixth term = (4 x 59733) + 1 = 238933.

||||||||

{ 772 }

Solve this: what comes next in this sequence?:

111, 144, 177, 210, 243, ...?

||||

By taking the differences between adjacent terms, we can see that this is an Arithmetic Progression, with common difference = 33. Therefore, the sixth term can be calculated by adding the common difference to the fifth term. It is equal to 243 + 33 = 276.

|||||||||

{ 773 }

Which number fits in the next spot in this sequence?:

8.1, 16.2, 32.4, 64.8, 129.6, ...?

||||

By taking the ratios of adjacent terms, we can see that this is a Geometric Progression, with common ratio = 2. Therefore, the sixth term can be calculated by multiplying the fifth term with the common ratio. It is equal to 129.6 x 2 = 259.2.

|||||||||

{ 774 }

Solve this: what comes next in this sequence?:

294, 344, 394, 444, 494, ...?

||||

By taking the differences between adjacent terms, we can see that this is an Arithmetic Progression, with common difference = 50. Therefore, the sixth term can be calculated by adding the common difference to the fifth

term. It is equal to 494 + 50 = 544.

||||||||

{ 775 }

Which number fits in the next spot in this sequence?:

234, 469, 939, 1879, 3759, ...?

||||

By observation and trial and error, we find that each term in this

sequence is obtained by multiplying the previous term by 2 and adding 1.

Therefore, the sixth term = (2 x 3759) + 1 = 7519.

||||||||

{ 776 }

Solve this: what comes next in this sequence?:

154, 191, 228, 265, 302, ...?

||||

By taking the differences between adjacent terms, we can see that this is

an Arithmetic Progression, with common difference = 37. Therefore, the

sixth term can be calculated by adding the common difference to the fifth

term. It is equal to 302 + 37 = 339.

||||||||

{ 777 }

Which number fits in the next spot in this sequence?:

264, 1057, 4229, 16917, 67669, ...?

||||

By observation and trial and error, we find that each term in this sequence is obtained by multiplying the previous term by 4 and adding 1. Therefore, the sixth term = (4 x 67669) + 1 = 270677.

|||||||||

{ 778 }

Which number fits in the next spot in this sequence?:

124, 497, 1989, 7957, 31829, ...?

||||

By observation and trial and error, we find that each term in this sequence is obtained by multiplying the previous term by 4 and adding 1. Therefore, the sixth term = (4 x 31829) + 1 = 127317.

|||||||||

{ 779 }

Which number fits in the next spot in this sequence?:

238, 953, 3813, 15253, 61013, ...?

||||

By observation and trial and error, we find that each term in this

sequence is obtained by multiplying the previous term by 4 and adding 1.

Therefore, the sixth term = (4 x 61013) + 1 = 244053.

||||||||

{ 780 }

Which number fits in the next spot in this sequence?:

2.1, 6.3, 18.9, 56.7, 170.1, ...?

||||

By taking the ratios of adjacent terms, we can see that this is a

Geometric Progression, with common ratio = 3. Therefore, the sixth term

can be calculated by multiplying the fifth term with the common ratio. It is

equal to 170.1 x 3 = 510.3.

||||||||

{ 781 }

Solve this: what comes next in this sequence?:

291, 323, 355, 387, 419, ...?

||||

By taking the differences between adjacent terms, we can see that this is

an Arithmetic Progression, with common difference = 32. Therefore, the

sixth term can be calculated by adding the common difference to the fifth

term. It is equal to 419 + 32 = 451.

||||||||

{ 782 }

Which number fits in the next spot in this sequence?:

285, 856, 2569, 7708, 23125, ...?

||||

By observation and trial and error, we find that each term in this

sequence is obtained by multiplying the previous term by 3 and adding 1.

Therefore, the sixth term = (3 x 23125) + 1 = 69376.

|||||||||

{ 783 }

Which number fits in the next spot in this sequence?:

94, 283, 850, 2551, 7654, ...?

||||

By observation and trial and error, we find that each term in this

sequence is obtained by multiplying the previous term by 3 and adding 1.

Therefore, the sixth term = (3 x 7654) + 1 = 22963.

|||||||||

{ 784 }

Which number fits in the next spot in this sequence?:

8.8, 17.6, 35.2, 70.4, 140.8, ...?

||||

By taking the ratios of adjacent terms, we can see that this is a

Geometric Progression, with common ratio = 2. Therefore, the sixth term can be calculated by multiplying the fifth term with the common ratio. It is equal to 140.8 x 2 = 281.6.

||||||||

{ 785 }

Solve this: what comes next in this sequence?:

193, 218, 243, 268, 293, ...?

||||

By taking the differences between adjacent terms, we can see that this is an Arithmetic Progression, with common difference = 25. Therefore, the sixth term can be calculated by adding the common difference to the fifth term. It is equal to 293 + 25 = 318.

|||||||||

{ 786 }

Solve this: what comes next in this sequence?:

138, 174, 210, 246, 282, ...?

||||

By taking the differences between adjacent terms, we can see that this is an Arithmetic Progression, with common difference = 36. Therefore, the sixth term can be calculated by adding the common difference to the fifth term. It is equal to 282 + 36 = 318.

||||||||

{ 787 }

Which number fits in the next spot in this sequence?:

2.2, 8.8, 35.2, 140.8, 563.2, ...?

||||

By taking the ratios of adjacent terms, we can see that this is a

Geometric Progression, with common ratio = 4. Therefore, the sixth term

can be calculated by multiplying the fifth term with the common ratio. It is

equal to 563.2 x 4 = 2252.8.

||||||||

{ 788 }

Which number fits in the next spot in this sequence?:

260, 781, 2344, 7033, 21100, ...?

||||

By observation and trial and error, we find that each term in this

sequence is obtained by multiplying the previous term by 3 and adding 1.

Therefore, the sixth term = (3 x 21100) + 1 = 63301.

||||||||

{ 789 }

Solve this: what comes next in this sequence?:

100, 119, 138, 157, 176, ...?

||||

By taking the differences between adjacent terms, we can see that this is

an Arithmetic Progression, with common difference = 19. Therefore, the

sixth term can be calculated by adding the common difference to the fifth

term. It is equal to 176 + 19 = 195.

|||||||||

{ 790 }

Solve this: what comes next in this sequence?:

223, 237, 251, 265, 279, ...?

||||

By taking the differences between adjacent terms, we can see that this is

an Arithmetic Progression, with common difference = 14. Therefore, the

sixth term can be calculated by adding the common difference to the fifth

term. It is equal to 279 + 14 = 293.

|||||||||

{ 791 }

Which number fits in the next spot in this sequence?:

263, 790, 2371, 7114, 21343, ...?

||||

By observation and trial and error, we find that each term in this sequence is obtained by multiplying the previous term by 3 and adding 1. Therefore, the sixth term = (3 x 21343) + 1 = 64030.

|||||||||

{ 792 }

Which number fits in the next spot in this sequence?:

4.8, 19.2, 76.8, 307.2, 1228.8, ...?

||||

By taking the ratios of adjacent terms, we can see that this is a Geometric Progression, with common ratio = 4. Therefore, the sixth term can be calculated by multiplying the fifth term with the common ratio. It is equal to 1228.8 x 4 = 4915.2.

|||||||||

{ 793 }

Which number fits in the next spot in this sequence?:

4.9, 19.6, 78.4, 313.6, 1254.4, ...?

||||

By taking the ratios of adjacent terms, we can see that this is a Geometric Progression, with common ratio = 4. Therefore, the sixth term can be calculated by multiplying the fifth term with the common ratio. It is

equal to 1254.4 x 4 = 5017.6.

||||||||

{ 794 }

Solve this: what comes next in this sequence?:

196, 215, 234, 253, 272, ...?

||||

By taking the differences between adjacent terms, we can see that this is

an Arithmetic Progression, with common difference = 19. Therefore, the

sixth term can be calculated by adding the common difference to the fifth

term. It is equal to 272 + 19 = 291.

|||||||||

{ 795 }

Solve this: what comes next in this sequence?:

232, 270, 308, 346, 384, ...?

||||

By taking the differences between adjacent terms, we can see that this is

an Arithmetic Progression, with common difference = 38. Therefore, the

sixth term can be calculated by adding the common difference to the fifth

term. It is equal to 384 + 38 = 422.

|||||||||

{ 796 }

Solve this: what comes next in this sequence?:

94, 144, 194, 244, 294, ...?

||||

By taking the differences between adjacent terms, we can see that this is an Arithmetic Progression, with common difference = 50. Therefore, the sixth term can be calculated by adding the common difference to the fifth term. It is equal to 294 + 50 = 344.

|||||||||

{ 797 }

Solve this: what comes next in this sequence?:

204, 217, 230, 243, 256, ...?

||||

By taking the differences between adjacent terms, we can see that this is an Arithmetic Progression, with common difference = 13. Therefore, the sixth term can be calculated by adding the common difference to the fifth term. It is equal to 256 + 13 = 269.

|||||||||

{ 798 }

Which number fits in the next spot in this sequence?:

8.4, 16.8, 33.6, 67.2, 134.4, ...?

||||

By taking the ratios of adjacent terms, we can see that this is a

Geometric Progression, with common ratio = 2. Therefore, the sixth term

can be calculated by multiplying the fifth term with the common ratio. It is

equal to 134.4 x 2 = 268.8.

||||||||

{ 799 }

Which number fits in the next spot in this sequence?:

175, 351, 703, 1407, 2815, ...?

||||

By observation and trial and error, we find that each term in this

sequence is obtained by multiplying the previous term by 2 and adding 1.

Therefore, the sixth term = (2 x 2815) + 1 = 5631.

||||||||

{ 800 }

Solve this: what comes next in this sequence?:

178, 228, 278, 328, 378, ...?

||||

By taking the differences between adjacent terms, we can see that this is

an Arithmetic Progression, with common difference = 50. Therefore, the

sixth term can be calculated by adding the common difference to the fifth

term. It is equal to 378 + 50 = 428.

||||||||

{ 801 }

Which number fits in the next spot in this sequence?:

2.7, 10.8, 43.2, 172.8, 691.2, ...?

||||

By taking the ratios of adjacent terms, we can see that this is a

Geometric Progression, with common ratio = 4. Therefore, the sixth term

can be calculated by multiplying the fifth term with the common ratio. It is

equal to 691.2 x 4 = 2764.8.

||||||||

{ 802 }

Solve this: what comes next in this sequence?:

139, 160, 181, 202, 223, ...?

||||

By taking the differences between adjacent terms, we can see that this is

an Arithmetic Progression, with common difference = 21. Therefore, the

sixth term can be calculated by adding the common difference to the fifth

term. It is equal to 223 + 21 = 244.

||||||||

{ 803 }

Which number fits in the next spot in this sequence?:

3.3, 13.2, 52.8, 211.2, 844.8, ...?

||||

By taking the ratios of adjacent terms, we can see that this is a

Geometric Progression, with common ratio = 4. Therefore, the sixth term

can be calculated by multiplying the fifth term with the common ratio. It is

equal to 844.8 x 4 = 3379.2.

|||||||||

{ 804 }

Which number fits in the next spot in this sequence?:

1.3, 5.2, 20.8, 83.2, 332.8, ...?

||||

By taking the ratios of adjacent terms, we can see that this is a

Geometric Progression, with common ratio = 4. Therefore, the sixth term

can be calculated by multiplying the fifth term with the common ratio. It is

equal to 332.8 x 4 = 1331.2.

|||||||||

{ 805 }

Which number fits in the next spot in this sequence?:

9.2, 36.8, 147.2, 588.8, 2355.2, ...?

||||

By taking the ratios of adjacent terms, we can see that this is a

Geometric Progression, with common ratio = 4. Therefore, the sixth term

can be calculated by multiplying the fifth term with the common ratio. It is

equal to 2355.2 x 4 = 9420.8.

|||||||||

{ 806 }

Which number fits in the next spot in this sequence?:

114, 343, 1030, 3091, 9274, ...?

||||

By observation and trial and error, we find that each term in this

sequence is obtained by multiplying the previous term by 3 and adding 1.

Therefore, the sixth term = (3 x 9274) + 1 = 27823.

|||||||||

{ 807 }

Which number fits in the next spot in this sequence?:

1.1, 3.3, 9.9, 29.7, 89.1000000000001, ...?

||||

By taking the ratios of adjacent terms, we can see that this is a

Geometric Progression, with common ratio = 3. Therefore, the sixth term

can be calculated by multiplying the fifth term with the common ratio. It is

equal to 89.1000000000001 x 3 = 267.3.

||||||||

{ 808 }

Which number fits in the next spot in this sequence?:

196, 393, 787, 1575, 3151, ...?

||||

By observation and trial and error, we find that each term in this

sequence is obtained by multiplying the previous term by 2 and adding 1.

Therefore, the sixth term = (2 x 3151) + 1 = 6303.

|||||||||

{ 809 }

Which number fits in the next spot in this sequence?:

221, 664, 1993, 5980, 17941, ...?

||||

By observation and trial and error, we find that each term in this

sequence is obtained by multiplying the previous term by 3 and adding 1.

Therefore, the sixth term = (3 x 17941) + 1 = 53824.

|||||||||

{ 810 }

Which number fits in the next spot in this sequence?:

217, 652, 1957, 5872, 17617, ...?

||||

By observation and trial and error, we find that each term in this

sequence is obtained by multiplying the previous term by 3 and adding 1.

Therefore, the sixth term = (3 x 17617) + 1 = 52852.

|||||||||

{ 811 }

Which number fits in the next spot in this sequence?:

253, 507, 1015, 2031, 4063, ...?

||||

By observation and trial and error, we find that each term in this

sequence is obtained by multiplying the previous term by 2 and adding 1.

Therefore, the sixth term = (2 x 4063) + 1 = 8127.

|||||||||

{ 812 }

Solve this: what comes next in this sequence?:

181, 223, 265, 307, 349, ...?

||||

By taking the differences between adjacent terms, we can see that this is

an Arithmetic Progression, with common difference = 42. Therefore, the sixth term can be calculated by adding the common difference to the fifth term. It is equal to 349 + 42 = 391.

IIIIIIII

{ 813 }

Which number fits in the next spot in this sequence?:

92, 185, 371, 743, 1487, ...?

IIII

By observation and trial and error, we find that each term in this sequence is obtained by multiplying the previous term by 2 and adding 1. Therefore, the sixth term = (2 x 1487) + 1 = 2975.

IIIIIIII

{ 814 }

Solve this: what comes next in this sequence?:

204, 231, 258, 285, 312, ...?

IIII

By taking the differences between adjacent terms, we can see that this is an Arithmetic Progression, with common difference = 27. Therefore, the sixth term can be calculated by adding the common difference to the fifth term. It is equal to 312 + 27 = 339.

IIIIIIII

{ 815 }

Solve this: what comes next in this sequence?:

197, 214, 231, 248, 265, ...?

IIII

By taking the differences between adjacent terms, we can see that this is an Arithmetic Progression, with common difference = 17. Therefore, the sixth term can be calculated by adding the common difference to the fifth term. It is equal to 265 + 17 = 282.

IIIIIIII

{ 816 }

Which number fits in the next spot in this sequence?:

198, 595, 1786, 5359, 16078, ...?

IIII

By observation and trial and error, we find that each term in this sequence is obtained by multiplying the previous term by 3 and adding 1. Therefore, the sixth term = (3 x 16078) + 1 = 48235.

IIIIIIII

{ 817 }

Which number fits in the next spot in this sequence?:

5.7, 22.8, 91.2, 364.8, 1459.2, ...?

IIII

By taking the ratios of adjacent terms, we can see that this is a Geometric Progression, with common ratio = 4. Therefore, the sixth term can be calculated by multiplying the fifth term with the common ratio. It is equal to 1459.2 x 4 = 5836.8.

|||||||||

{ 818 }

Solve this: what comes next in this sequence?:

267, 315, 363, 411, 459, ...?

||||

By taking the differences between adjacent terms, we can see that this is an Arithmetic Progression, with common difference = 48. Therefore, the sixth term can be calculated by adding the common difference to the fifth term. It is equal to 459 + 48 = 507.

|||||||||

{ 819 }

Which number fits in the next spot in this sequence?:

5.8, 17.4, 52.2, 156.6, 469.8, ...?

||||

By taking the ratios of adjacent terms, we can see that this is a Geometric Progression, with common ratio = 3. Therefore, the sixth term can be calculated by multiplying the fifth term with the common ratio. It is

equal to 469.8 x 3 = 1409.4.

||||||||

{ 820 }

Solve this: what comes next in this sequence?:

91, 115, 139, 163, 187, ...?

||||

By taking the differences between adjacent terms, we can see that this is an Arithmetic Progression, with common difference = 24. Therefore, the sixth term can be calculated by adding the common difference to the fifth term. It is equal to 187 + 24 = 211.

||||||||